LIFE EVERYWHERE

LIFE
EVERYWHERE

*The Maverick Science
of Astrobiology*

DAVID DARLING

BASIC
BOOKS

A Member of the Perseus Books Group

Published by Basic Books,
A Member of the Perseus Books Group

ISBN: 0-465-01563-8
First Edition

Library of Congress Cataloging-in-Publication Data

Darling, David J.
Life everywhere : the maverick science of astrobiology / by David Darling.
p. cm.
Includes bibliographical references and index.
ISBN 0-465-01563-8 (alk. paper)
1. Exobiology. 2. Life—Origin. I. Title.

QH327 .D37 2001
576.8'39—dc21

2001025147

The paper used in this publication meets the requirements of the
American National Standard for Permanence of Paper for Printed Library
Materials Z39.48-1984.

01 02 03 04 / 10 9 8 7 6 5 4 3 2 1

Designed by Nighthawk Design

With Love,
to my wife, Jill
my children, Lori-An and Jeff
and my parents, Eric and Marjorie Darling

CONTENTS

ACKNOWLEDGMENTS

First, I'd like to thank Bill Frucht, senior editor at Basic Books, who played an essential role in the genesis and development of this book. To him goes credit for the original concept, much of the organization, and numerous intelligent and insightful suggestions.

For providing me with information, opinions and advice, I'm grateful in particular to: Gustaf Arrhenius, Scripps Institution of Oceanography; Mark Bedau, Reed College, Portland, Oregon; Max Bernstein, NASA Ames Research Center; Jay Brandes, University of Texas at Austin; Simon Conway Morris, University of Cambridge; Alex Ellery, Queen Mary and Westfield College, London; Clark Friend, Oxford Brookes University; Guillermo Gonzalez, University of Washington; Joseph Kirschvinck, California Institute of Technology; Geoffrey Marcy, University of California, Berkeley; Stephen Mojzsis, University of California, Los Angeles; Thomas Ray, University of Oklahoma; Gregory Schmidt, NASA Ames Research Center; Victor Tejfel, Fesenkov Astrophysical Institute; Hojatollah Vali, McGill University; and Peter Ward, University of Washington, Seattle. Any mistakes or misinterpretations that may have crept into the text are, of course, my responsibility alone.

As always, my agent Patricia van der Leun worked tirelessly behind the scenes and provided support above and beyond the call of duty. Thank you again, Patricia.

Last, and most of all, I thank my family. Only through their support and encouragement, over many years, have I been lucky enough to pursue a fantasy career instead of a "real job."

PREFACE

Something extraordinary has happened over the past decade. Without any fanfare, scientists the world over have reached a consensus on one of the most profound questions ever to challenge the human mind: Are we alone? In all of this vast and ancient cosmos, is life confined to Earth?

No. Almost beyond doubt, life exists elsewhere. Probably, in microbial form at least, it is widespread. And more likely than not, we will find incontestable evidence of it quite soon—perhaps within the next ten to twenty years. These are the core elements of the remarkable new accord that is now routinely accepted by researchers across a spectrum of disciplines.

Behind this surge in scientific optimism about the prospects for alien life lies a rush of remarkable discoveries. A bewildering assortment of (mostly microscopic) life-forms has been found thriving in what were once thought to be uninhabitable regions of our planet. These hardy creatures have turned up in deep, hot underground rocks, around scalding volcanic vents at the bottom of the ocean, in the desiccated, super-cold Dry Valleys of Antarctica, in places of high acid, alkaline, and salt content, and below many meters of polar ice. The range of locales where organisms could be expected to survive in the universe is thus vastly expanded. Some deep-dwelling, heat-loving microbes, genetic studies suggest, are among the oldest species known, hinting that not only can life thrive indefinitely in what appear to us totally alien environments, it may actually *originate* in such places. If so—if the cradles of biogenesis tend to be hot, dark, subsurface hells rather than our familiar sun-drenched surface edens—then the widespread appearance of life throughout the cosmos is made much more likely.

Scientific opinion has also shifted dramatically toward the view that life may be "easy"—able to assemble itself from simpler components at the slightest opportunity. How else to account for the signs in ancient rocks that bacteria proliferated on Earth as long ago as 3.8 billion years, during the intense bombardment phase following the birth of our solar system? Terrestrial life appeared almost before it had a reasonable chance of long-term survival, then somehow managed to weather the ferocious early storm of asteroid and comet impacts. An increasingly common claim among researchers is that life may arise inevitably whenever a suitable energy source, a concentrated supply of organic (carbon-based) material, and water occur together.

These ingredients are starting to look ubiquitous in space. Comets, in particular, are increasingly seen as significant vehicles for delivering water and organic cocktails to infant worlds. And with the discovery on Earth of meteorites from Mars, the *interplanetary* transfer of biochemicals or even life itself has become a respectable topic of debate.

Both within and beyond the solar system, the list of potential places where life may have become established is growing fast. Close to home, Jupiter's moons Europa and Ganymede have taken on biological interest with the realization that they may harbor chemical-rich oceans of water beneath their icy surfaces. Farther afield, the finding of dozens of extrasolar planets— almost as soon as we knew how to look for them—encourages scientists to think that planetary systems around stars are the rule rather than the exception. From origin of life studies to complexity theory, from extrasolar planet detection to work on extremophiles, from pre-Cambrian paleontology to interstellar chemistry, the emerging message is clear and virtually unanimous: extraterrestrial life is there for the finding.

Having generally agreed that it's only a matter of time before the first alien organisms come to light, scientists are now busily laying the foundations of the new field of astrobiology.* They are asking: What are the general conditions needed for life to appear? How common will it prove to be? Where will it be found? What will it be like? These are now respectable and intensely debated mainstream issues—the subject of major conferences and numerous scientific papers every month, a focal point of interdepartmental projects at a growing number of universities and other research organizations, and a key motivating influence behind international space programs. They are the raison d'être of NASA's Astrobiology Institute, which began operations in 1998 and saw the Nobel Prize winner Baruch Blumberg (renowned

---

*The study of life in the universe has been variously called exobiology (first by geneticist Joshua Lederberg in 1960), bioastronomy (a more recent name), and astrobiology (oldest of all, mentioned in Soviet literature as far back as 1953). These terms are still used more or less interchangeably, but "astrobiology" is in the ascendancy thanks to its recent adoption by NASA.

for his work on hepatitis B) appointed as its head in 1999. They motivate the University of Washington's astrobiology program in Seattle, which welcomed its first intake of graduate students in 1999. They formed the topic of the first annual conference devoted to the science of astrobiology, held at NASA Ames in April 2000. "We are witnessing not just a shift in scientific paradigm but, more important, a shift in cultural acceptability among scientists," said extrasolar planet hunter Geoff Marcy. Poised on the brink of a momentous breakthrough that will change forever how humankind thinks about itself and the universe around it, astrobiology is quickly coming of age.

Like all branches of science reaching toward maturity, astrobiology is alive with diverse theories, experimental data, rumors and conjecture. These are extraordinarily frenetic times, an immensely fertile period for thought. The corridors outside conference halls are crowded with researchers in animated conversation, vying to get across their points of view, forming camps of opinion, pushing back the frontiers of knowledge and surmising what lies out there.

At first glance, it may seem that apart from their broad agreement that terrestrial life is not unique, those engaged in this new endeavor are more in discord than harmony. Certainly there are many differences of opinion about specific issues, from claims about Martian "fossils" to the steps involved in life's genesis. That's to be expected. But it isn't too early to make out, amid the tumult of claims and counterclaims, the beginnings of a *general theory of biology*, a framework of concepts that underpins the development of life wherever it takes place.

This book is a report from the frontline of astrobiological research, an examination of the issues, arguments and experimental results foremost in the minds of those who are spearheading this astonishing new field. Beyond that, it is an attempt to see the way ahead, to identify the concepts that may eventually unify our understanding of life in a broader context. On what may be the brink of our first encounter with an alien species, we ask: What principles govern the emergence and evolution of life throughout the cosmos? Where can we expect to find other living worlds, and what will we discover on them?

1

The Intimate Mystery

Nothing could be more familiar than life. But what exactly *is* it? On a practical level, how can we tell life from nonlife wherever it occurs in the universe?

Defining life hasn't traditionally been the biologist's favorite pursuit. The English geneticist J. B. S. Haldane began his 1947 essay "What is Life?" with the statement: "I am not going to answer this question." Scientists don't need a dictionary to tell them that a field of daffodils or a colony of bacteria is alive and a tailor's dummy isn't. Biology has gotten along quite nicely without specifically saying what it's studying. But astrobiology doesn't have that luxury. How can we hope to find life on other worlds if we don't know what we're looking for?

Maybe we'll be lucky. When future probes melt their way through the icy coating of Jupiter's moon Europa, they may send back glimpses of giant luminous creatures patrolling a Stygian sea. When the first manned expedition to Mars samples the bed of the ancient ocean that once sprawled across the northern hemisphere, it may unearth the perfectly preserved fossil of a Martian trilobite. The late Carl Sagan was among those who suggested that something big might lumber before the watching cameras of *Viking* on Mars or float visibly in the cloud-tops of Jupiter as the *Voyager* probes flew by.

Recognizing such large and obvious extraterrestrial life (or its remains) would be child's play. But the universe isn't likely to be so accommodating. Life may only rarely crop up on a grand scale. It could also be utterly bizarre, unlike anything we've previously met or imagined. And even if it follows a more familiar pattern, confirming its presence from far away will hinge on our ability to distinguish, clearly and unambiguously, the true signatures of biological activity.

☼

So what is life exactly? "Something that can make copies of itself," according to a familiar textbook definition. That would certainly include every organism on Earth. Even in special cases, like those of mules, celibates, and men who have undergone vasectomies, where the individual can't or chooses not to engage in procreation, the Xeroxing of DNA goes on all the time at the cellular level.

For many scientists, however, while self-replication is a necessary feature of living things, it isn't the most fundamental. Stanley Miller, a biochemist at the University of California, San Diego who did some of the pioneering experiments on the chemical origin of life, makes no bones about his dislike of definitions: "[They] are what you impose on your thoughts. There are so many more important things to discover that to engage in an extended discussion over definitions, I think, is a waste of time." Having said this, his money is firmly on *evolution* as the sine qua non of life. "My definition of life, viewed from the perspective of origins, is that the origin of life is the origin of evolution." Evolution in turn involves three key factors: replication, selection, and mutation. "Replication is the hard part. Selection is where nature selects out the ones that would replicate the fastest, and mutation means that you make a small number of errors. It is important that those mutations, or errors, be propagated on to the progeny, so the organism improves. Reproduction is simply making an accurate copy of genetic material."

Another origin-of-life researcher, Antonio Lazcano at the National Autonomous University of Mexico, holds a similar view: "Some people would say that as long as you have a single molecule that is able to replicate and evolve, that is enough. My own tendency is to define life as a system that is able to undergo Darwinian evolution. By this, I mean a chemical system that can actually undergo a process of mutations and rearrangement of the genetic material, and can adapt to the environment." For Lazcano, as for Miller, "the questions of defining life and the origins of life are connected."

Mark Bedau, a philosopher of biology at Reed College in Portland, Oregon, goes a step further. He regards evolution as "the thing which explains why all the other properties are there—the essence, the root cause." What's really alive is the whole system: an ensemble of countless individual organisms of many species, all interacting, reproducing, and displaying unpredictable, open-ended evolution.

This is an idea with far-reaching implications. If life can be *anything* that shows open-ended evolution, then it isn't locked into a particular material form. It doesn't have to be carbon-based. In principle, it doesn't have to be chemically based at all. And if that sounds too far-fetched to take seriously, then watch out. Wildly unfamiliar creatures are already lurking in laboratories in the United States, Japan, Italy, Germany, Britain and elsewhere, and have even gained access to the Internet. They don't look like us. Their origin is completely different from that of any natural organism on Earth. In essence, they inhabit an alternative stratum of reality. Yet there they are, breeding, growing, competing, dying, evolving, just like the rest of us. They are *artificial* life-forms—"a-life"—and their home is the digital landscape of computers.

Thomas Ray, a professor of zoology at the University of Oklahoma, is one of the pioneering investigators of these new, nonorganic organisms and author of the Tierra a-life software system. Genesis inside Tierra dawns with a single, minuscule progenitor, the "Ancestor." It's a tiny string of machine code, just 80 bytes long, brought into existence with the capacity to make copies of itself inside the computer's working memory. The Ancestor spawns a daughter program. Then Ancestor and daughter each replicate again, as do their offspring, and so it goes on, multiplication upon multiplication. The little programs, with their self-copying ability, are simple analogues of the nucleic-acid-based genetic code of biological life. And crucially, just like that DNA-mediated system, Ray's self-replicators are slightly less than perfect. They don't always result in exact copies of the original because the Tierra environment is set up so as to occasionally reach in and randomly flip one of the bits—the binary digits—in a daughter program, making it genetically distinct from its parent. Usually the switch is bad news, rendering a program unable to copy itself as well as before, if at all. But sometimes turning a zero into a one or vice versa works to the creature's advantage, enabling it to multiply a little faster than its rivals. In this way, mutation, the master key to novelty and adaptation, is introduced into the proceedings. By the time the computer's memory is chockfull of Tierrans, there are all manner of variations on the original theme—a host of genetically distinct self-copiers battling for survival in their overcrowded electronic domain. At this point the real fun begins. In accordance with certain "fitness" criteria built into the system at the outset, the little programs begin competing for memory space. The success of a particular species, or byte-string, depends on how effectively it can replicate and transmit its genes to the next generation, or even usurp its rivals' private memory space.

Ray believes that systems like his provide the first experimental basis for a *comparative* biology:

> Life on Earth is the product of evolution by natural selection operating in the medium of carbon chemistry. However, in theory, the process of evolution is neither limited to occurring on the Earth, nor in carbon chemistry. Just as it may occur on other planets, it may also operate in other media, such as the medium of digital computation. And just as evolution on other planets is not a model of life on Earth, nor is natural evolution in the digital medium.

Like others in his field, Ray is adamant that his creations are "not models of life but independent instances of life." Forget semantics. Forget metaphysical musings on the meaning or nature of life. If it acts like a duck, it's a duck; if it evolves, it's alive. This is proof-of-pudding empiricism like that employed by an old chestnut in the field of artificial intelligence, the Turing Test. If by questioning alone, says the Turing Test, you can't tell which of two interviewees is human and which is machine, then the machine should be considered to be genuinely intelligent. But whereas the Turing Test is a long way from being applied in practice, the "strong" a-life claim is with us here and now—and it is extraordinarily radical. Life, it suggests, can be defined without reference to a material medium. Its fundamental essence isn't solid, liquid, or gas, or any kind of chemistry, or even a digital dance of electrons. The form of matter is irrelevant. What distinguishes life, at its most basic level, is *information*.

That's a lot for flesh-and-blood bipeds to swallow. We may live and pass on life courtesy of the encyclopedic database enshrined within our DNA, but like every other terrestrial organism, we depend upon our carbonaceous bodies. We're material beings. The point the strong a-life claim makes, however, is not that life can exist in the absence of a medium (chemical or otherwise), but that the medium isn't what matters. What does is that there are general principles of the living state that are *independent* of a particular implementation. These principles, the idea goes, operate purely at the level of the informational and organizational substructure of life. Consequently, they apply anywhere in the universe.

In physics, this kind of abstraction is routine. Physicists make a living out of searching for—and finding—relationships that underpin otherwise seemingly diverse phenomena. Since at least the time of Galileo, the inorganic world has been well known to have a mathematical infrastructure. But we're not used to thinking about life in such terms. The theoretical physicist may be happy to work in a world of equations, a Platonic universe that stands behind the reality we perceive. But it comes as a shock to be told that a similar, intangible domain of symbols and logical relationships may form the backdrop to the very phenomenon of life. Obviously we're more than mere dust-devils of data. Still, as one of the originators of the a-life field, Christopher Langton, put it:

> There's nothing implicit about the material of anything—if you can capture its logical organization in some other medium you can have that same "machine," because it's the organization that constitutes the machine, not the stuff it's made of.

A-life researchers aren't the first to make this argument. The naturalist D'Arcy Wentworth Thompson was drawing attention to the common mathematical architecture of organisms in 1917 in his magnum opus, *On Growth and Form*. More recently, the Chilean biologists Humberto Maturana and Francisco Varela have sought to build a theoretical foundation for life in its broadest sense in terms of autopoiesis ("self-creation")—the ability of a system to invent and define itself by virtue of having a circular organization. The cells of terrestrial organisms, for instance, are autopoietic because they're made up of a physically bounded network of chemicals that, through an intricate series of reactions, actually generates the very network, together with the boundary—the membrane—that sets the system apart from its surroundings.

Autopoiesis is an all or nothing affair, because unless a system has a closed organization (albeit it may have an open *structure* to allow the inflow and outflow of energy and materials), it can't manufacture itself from within. Which prompts the question: How could such a completely self-reflexive entity get off the ground in the first place? How could life originate if it had to, as it were, pull itself up by its own bootstraps? As Stuart Kauffman, at the Santa Fe Institute in New Mexico, sees it:

> Life emerged . . . not simple, but complex and whole, and has remained complex and whole ever since—not because of a mysterious *élan vital*, but thanks to the simple, profound transformation of dead molecules into an organization by which each molecule's formation is catalyzed by some other molecule in the organization.

Kauffman doesn't go as far as those in the a-life field, who'd like to ground the definition of life in purely abstract and logical terms. He's orthodox to the extent that he regards every living thing as having both a chemical aspect (a body) and an informational aspect (a genome), and believes that it's meaningless to talk about one without the other. But he thinks a new concept, that of an "autonomous agent," is needed as the unifying factor of life. "It's not matter, it's not energy, it's not information," he says. "It's something else."

✵

Such sweeping theories of life promise to do for biology what Newton's theory of gravity did for physics—form the basis of a science whose principles apply everywhere in the cosmos. But right now astrobiologists have a more pressing concern. They simply want to *detect* extraterrestrial life. A single example—any example—will do. It might be a beleaguered underground colony of Martian microbes, a battered, desiccated cell scraped from some frozen ice-field on Callisto, or a radiation-seared viroid particle blown in on the stellar wind. The overwhelming priority for this raw young science is to find *something* that is biogenic and didn't come from Earth. That means focusing on a property of life that will give a reading on an instrument—that will actually generate a signal indicating that a living process is (or has been) at work. "When you think about how you recognize life," says Michael Meyer, head of NASA's astrobiology program in Washington, D.C., "you think about what life does, not what it is."

One of the things life does—that underpins its very existence—is metabolize. Biologists and astrobiologists are unanimous that metabolism has to be a linchpin of life everywhere. The term comes from the Greek *metabole*, meaning "change," and it refers to the functions and effects of the suite of interlocking chemical reactions found within an organism. A pivotal aspect of metabolism is the harnessing of energy. Living things, whatever their nature, must be able to capture energy from their surroundings, turn it into a storable form, and then release it when needed in precisely controlled amounts. Energy must be available on demand for essential biological tasks such as building complex substances from simpler starting materials, effecting repairs to living structures, and reproducing.

These processes require that certain chemicals be taken in from the surroundings and certain others given out. The particular chemicals involved and the specific way in which the chemical balance of the environment is affected are signs that a biological process is at work. And that fact is the crux as far as astrobiology is concerned. Metabolism causes changes that, given the knowledge and the right equipment, can be detected and used to diagnose the presence of life.

Even extinct organisms may leave clues to their metabolic activity—chemical and mineralogical traces in rocks that speak strongly or uniquely of a biogenic origin. Such rocks may arrive as meteorites splashed from the surface of neighboring planets following collisions with asteroids or, in the future, they may be brought back by sample-return probes. Close examination in the lab may reveal another familiar and possibly universal characteristic of life.

Every living thing on Earth exists inside some sort of bag. Those of us fortunate enough to be multicellular have a general outer wrapping—skin, scales, an exoskeleton, a waxy cuticle. But each individual cell, whether part of a larger organism or not, has its own bag, a cell membrane, that serves a variety of purposes. The metabolism of life-as-we-know-it requires that a rich cocktail of exotic chemicals be closely confined under conditions radically different from those in the immediate nonliving neighborhood.

On Earth, every organism, from the lowliest bacterium to a human being, stores energy in exactly the same way—in the form of the chemical bonds that link together the phosphate groups in the molecule known as adenosine triphosphate (ATP). Tacking on phosphate groups to make ATP stores energy, splitting them away sets energy free. This released energy drives all the other aspects of metabolism, including assembling new molecules such as proteins from simpler components (anabolism), breaking down existing large molecules (digestion, or catabolism), extracting chemical energy (respiration), and, if you happen to be a green plant, directly intercepting and making use of the energy in sunlight (photosynthesis).

A chart of the main metabolic pathways of terrestrial life reveals an interlocking network of reaction cycles and chains, marvelous in complexity and organization. But none of this chemical wizardry is self-starting. If all the reactants involved in metabolism were simply thrown together and kept roughly at room temperature, nothing would happen. The biochemical reactions upon which all known life depends just aren't energetically favorable; they don't take place merely by the various reactant molecules bumping into one another. They need to be helped along. The substances that do this, biological catalysts or *enzymes*, are mostly protein molecules (there are a few notable exceptions) and each one is specific to a particular reaction or small set of similar reactions, owing to its unique three-dimensional shape. This specificity is crucial because it means that by regulating the production of different enzymes, an organism can run a system of highly ordered, inter-linked reaction chains, rather than just a muddled-up chemical jamboree.

The result is a repeatable pattern of reactions that, without this contin-uous intervention, wouldn't occur or would quickly grind to a stop. In ther-modynamic terms, it's a system far out of chemical equilibrium. If this state doesn't quite qualify as a definition of life, it's certainly one of life's defining *characteristics*.

But enzymes, like many proteins, are delicate. If they get too hot, their shape changes, they start to fall apart, and they lose their catalytic ability. The same thing happens if their environment becomes too acid or alkaline. For these and other reasons, an important part of metabolism is *homeo-stasis*—the maintenance of a relatively stable ambience inside an organism. Whatever happens outside, it's crucial that an organism's internal state stay pretty much the same. The only way that can happen is if it exists within a kind of protective bubble. If life as we can reasonably imagine it and physico-chemical states that are far from equilibrium go hand in hand, then a means of containment and segregation is absolutely essential. And this is where cell membranes come in. The cell membrane holds the contents of an organism together and separates the region within which metabolism takes place from the outside world.

At the same time, the membrane isn't an impenetrable barrier. It's a subtly constructed interface, itself a product of metabolism, that allows, with the expenditure of some energy by its owner, the controlled two-way passage of certain substances that the organism needs both to acquire and dispose of. Could life exist without such an interface? Not according to Lynn Margulis, a biologist at the University of Massachusetts at Amherst, who has written much about life's nature and evolution. "Life," she says, "is a self-bounded system where the boundary is made by the material in the system. It's not a thing, it's a process, and these processes involve the production and mainte-nance of identity."

Jeffrey Bada, who directs NASA's astrobiology research at the Scripps Institution of Oceanography in La Jolla, California, disagrees. He has theo-rized that very primitive forms of life might exist as a sort of boundary-less broth. This may be how life started out on Earth, and it could be that the genesis of living things always involves an early precellular phase. If so, Bada argues, on worlds where there's little evolutionary pressure to force further change, life might never progress beyond its primordial soupy state. One such place, he suggests, might be the supposed underground ocean on Europa.

But if something like Bada's broth exists, would most scientists call it alive? Its discovery would certainly cause plenty of excitement. Not-quite-life, iffy-life, life-of-sorts—astrobiologists would gladly take whatever comes along, because anything remotely biological found on another world (assuming it had developed independently) would be powerful evidence that life is common throughout the universe. Still, the question strikes to the heart of how life is defined, especially at the lower end of the scale: Would an essentially unchanging sea of membrane-less, self-copying chemicals qualify as an instance of life? Probably not, in the judgment of most researchers. It would more likely be called *pre*biological. And the reason goes back to that key criterion of evolution. "The ability to evolve," insists Jack Szostak, a molecular biologist at the Massachusetts General Hospital in Boston, "is what distinguishes systems that are alive biologically from prebiotic chemical systems."

In the minds of most biologists, Darwinian-style evolution outranks replication, metabolism, or individuality as the chief definer of life. But these properties aren't mutually independent—quite the opposite. Evolution in the Darwinian sense *implies* that replication and natural selection are going on within a genetically diverse population of individuals. Replication implies metabolism. It's hard to see how these factors could be disentangled. As Oxford biologist John Maynard Smith put it, "entities with the properties of multiplication, variation, and heredity are alive, and entities lacking one or more of those properties are not."

If Darwinian evolution is what fundamentally marks out life from non-life or prelife then all living things must also, because of the way natural selection works, be capable of making copies of themselves. That implies the need for a genome—a complete set of instructions for self-replication. In theory, the genome *could* exist in any form. In Ray's Tierra system, it's a string of zeros and ones held in the computer's memory. But in nature, not only does the genome itself have to have a foot in the physical world (as a set of chromosomes, for example), but it also has to be encapsulated within a larger structure—a living organism. Some see this latter fact as almost incidental. The Oxford biologist Richard Dawkins has long championed the view that the gene—what he calls the "replicator"—is the central fact of life. In his view, cells and multicellular creatures evolved as mere vehicles to ensure the survival and transmission of their genetic cargo. That's pretty close to the a-life position, in which the emphasis is all on self-propagating patterns of information. But it's too extreme for most biologists, who tend to regard organisms as more than just housings for their genetic database. A more conventional view of life was suggested by the philosopher David Hull. In his scheme, biological evolution involves not only replicators (things that pass on their structure directly by replication), but also interactors (things that produce *differential* replication as a result of interacting as cohesive wholes with their environment) and lineages of these interactors. As he saw it:

> A process is a selection process because of the interplay between replication and interaction. The structure of replicators is differentially perpetuated because of the relative success of the interactors of which the replicators are part. In order to perform the functions they do, both replicators and interactors must be discrete individuals which come into existence and cease to exist. In this process they produce lineages which change indefinitely through time.

For terrestrial life, the replicators are genes, made from DNA, and the interactors are the many organisms found on Earth. On other worlds, the details of implementation may differ, but there seems every reason to suppose that the same overall arrangement applies to living things everywhere. According to Carl Emmeche, a philosopher who studies the nature of life at the Niels Bohr Institute in Copenhagen,

> It is highly conceivable that all life in the universe evolves by a kind of Darwinian selection of interactors, whose properties are in part specified by an informational storage that can be replicated. . . . [T]he very notion of natural selection and replication . . . seems to be specific for biological entities. . . . This definition is simple, elegant, general, and crystallizes our ideas of the general mechanism of the creation of living systems within an evolutionary perspective.

Again, it's possible to work with a paradigm of life like this on a very abstract level. Replicators, interactors, and lineages could all be set up inside a computer, for instance, or played with as patterns of little squares on gridded paper that follow certain rules. But astrobiologists, although they sometimes use computers for biological simulations, aren't so interested in such intangibilities. Their quest is for life in the *real* universe, life that has assembled itself from the raw materials on other worlds. And that brings us back to the other key factor of life—metabolism. Replicators and interactors can function in nature only if they have the means to manipulate energy and matter to their own ends. In the case of life that has evolved naturally, the informational and material aspects of life can't be divorced from one another. To both retain and act upon its onboard genome, for self-construction, self-maintenance, and reproduction, any living thing must harbor a network of component metabolites. That network, requiring conditions far out of equilibrium with the surroundings, can function only within some kind of boundary. And that boundary, in turn, defines an individual.

The nature of individuality might seem obvious. We think of ourselves as individuals. But in reality, each of us is a city—home to vast armies of bacteria encamped on the surface of our skin and mucous membranes, as well as within us—most of them, luckily for us, benign. More disturbingly, every cell of our bodies is inhabited profusely by beings from another place and time. *Mitochondria* are at the heart of energy production in the cell—the sites of cellular respiration. We literally can't lift a finger without them. Yet they have their own inner membranes and DNA, strikingly similar to those of bacteria. This is the essential clue to their probable origin. According to the endosymbiotic theory (first proposed in 1885, cast in modern form by Lynn Margulis, and now widely accepted), mitochondria, together with the light-harvesting chloroplasts found in green plant cells, are descendants of ancient, free-living microbes. At some point, more than a billion years ago, they became incorporated within larger cells as part of a symbiotic liaison, and there they have remained ever since. Are the mitochondria the "real" individuals and each of us a kind of hive in which they collectively dwell? And are we, in turn, mere elements of a much larger superorganism?

The web of life leaves no creature in isolation. Most obviously, social insects, like ants and bees, simply die if cut off from their swarm. But on a wider scale it's true of us all. From *Staphylococcus* to *Homo sapiens*, we're minuscule parts of a stupendously complex, planet-wide system of interconnected organic and inorganic components—the biosphere. Self-regulating through a myriad of feedback loops, endlessly cycling biological staples such as carbon, nitrogen, and water, is *this* the true quantum of life? Some supporters of the controversial Gaia theory think so. The whole biosphere is the primary life-form, the collective product of a host of lesser beings. Perhaps Cèzanne conveyed it best with his depiction of an apple as part fruit, part Earth. Anyway, it's a sobering thought that we who habitually think of ourselves as free and self-sufficient may be more like cells in a giant planetary superorganism.

Not all Gaia theorists go so far as to say that the Earth is a living individual. In any case, what matters from the perspective of astrobiology is that biospheres offer another opportunity for detecting life. Just as the environment within an organism is well out of equilibrium with its surroundings, so biospheres are expected to give themselves away by their markedly "unnatural" appearance. The originator of the Gaia theory, British chemist James Love-lock, wrote of Earth's atmosphere: "Almost everything about its composition seems to violate the laws of chemistry. . . . The air we breathe . . . can only be an artifact maintained in a steady state far from chemical equilibrium by biological processes."

On other worlds, too, life may have altered conditions on a planet-wide scale. So a key strategy in astrobiology will be to look for any combination of constituents in an atmosphere that is well out of normal chemical balance—a suspiciously unstable mixture that only living metabolisms could maintain.

☼

The more radical Gaia interpretation of an entire planet as a single life-form makes one wonder how unusual life might be elsewhere. Over the years, scientists and science fiction writers have dreamed up an extraordinary cosmic menagerie. In *The Black Cloud,* cosmologist Fred Hoyle imagined an intelligent, self-propelled interstellar cloud that arrives in the solar system and wreaks havoc by blocking out the Sun's light. Its "brain," a complex network of widely spaced molecules, can be expanded and reconfigured at will, giving the creature stupendous mental powers. Aerospace engineer Robert Forward, building on an idea by SETI pioneer Frank Drake, wrote about the diminutive, high-density inhabitants of a neutron star. His "cheelans," made of nuclear matter, live out their lives a million times faster than human beings, see in the far ultraviolet, and communicate by strumming the crust of their unusual stellar home with their abdomens.

Could such wildly flamboyant creatures actually exist? As Fred Hoyle wrote in the preface to *The Black Cloud*, "there is very little here that could not conceivably happen." Life elsewhere could be so strange that if we base our expectations too rigidly on terrestrial standards we might even have trouble recognizing it. Astrobiologists are well aware they have no way yet of putting constraints on the outer limits of life. Having only one data point to work with, they're compelled to be open-minded. Maybe there *are* star-dwelling communities, interstellar behemoths, energy-based life-forms, and other exotica that would put the *Star Trek* universe to shame. But while such speculation is entertaining and the subject of many an informal discussion between scientists, it isn't a central issue in professional circles. The dominant question is where the search for extraterrestrial life should be focused here and now. And the answer is evident from every astrobiological program, underway or planned, in which there is a significant investment of funds and other resources. It's evident in the "Roadmap" drawn up by the Astrobiology Institute to help guide NASA's activities in this field. It's evident in the overwhelming majority of papers published on the subject of extraterrestrial life in leading scientific journals and in the proceedings of relevant conferences, such as the first annual science conference on astrobiology held at the Ames Research Center in April 2000. Most tellingly, it's evident in the design and implementation of the multimillion-dollar instruments that have been built, or are being built, to test for the presence of biological activity on other worlds. The approach adopted by the scientific community is simple, straightforward, and practical: *to look for the kind of life we know*, allowing for possible adaptations to different environments.

The kind of life we know is, first and foremost, based on carbon. "No other element comes close to forming such a diverse array of bonds," explains Jeff Bada. Carbon's closest analogue is silicon, and there's been no shortage of speculation about the possibility of silicon-based life over the past century or so. In 1893, the chemist James Emerson Reynolds used his inaugural address to the British Association for the Advancement of Science to point out that the heat stability of silicon compounds might allow life to exist at very high temperatures. Picking up on this idea in an article published the following year, H. G. Wells wrote: "One is startled towards fantastic imaginings by such a suggestion: visions of silicon-aluminium organisms—why not silicon-aluminium men at once?—wandering through an atmosphere of gaseous sulphur." Thirty years later, J. B. S. Haldane proposed that life might be found deep inside a planet based on partly molten silicates.

At first sight, silicon does seem a promising alternative to carbon. Like carbon, it's common in the universe, and much of its basic chemistry is similar. Just as carbon combines with four hydrogen atoms to form methane, silicon yields silane; silicates are analogues of carbonates; both elements form long chains in which they alternate with oxygen; and so on. But on closer examination, silicon's biological credentials become less convincing. The biggest stumbling block seems to be the extreme ease with which silicon combines with oxygen. Wherever astronomers have looked—in meteorites, in comets, in the interstellar medium, in the outer layers of cool stars—they've found molecules of *oxidized* silicon (silicon dioxide and silicates) but no evidence at all of substances that might serve as the building-blocks of a silicon biochemistry. The silicon analogues of hydrocarbons—long chains of hydrogen-silicon compounds—are nowhere to be found. And there's a further problem with silicon dioxide. When carbon is oxidized during respiration, it becomes the gas carbon dioxide—a waste material that's easy for a creature to dispose of. But silicon dioxide turns into a solid—a crystalline lattice—the instant it forms. To put it mildly, that poses a respiratory challenge.

This difficulty didn't faze Stanley Weisbaum in his SF classic *A Martian Odyssey*. Observing the unusual behavior of one of the indigenous life-forms, a scientist in the novel notes:

> Those bricks were its waste matter. . . . We're carbon, and our waste matter is carbon dioxide, and this thing is silicon, and its waste is silicon dioxide—silica. But silica is a solid, hence the bricks. And it builds itself in, and when it is covered, it moves over to a fresh place to start over.

The door may still be ajar to the possibility of silicon-based biology—and for other novel biologies for that matter. But the fact remains that carbon really has no serious rival in the minds of most researchers who are actively involved in seeking out extraterrestrial life. The major point of debate is how much the *details* of the carbon chemistry of life will vary from one world to the next. Do all living things, for example, use DNA as their genetic material? Are the chemical pathways of their metabolism essentially the same? The Harvard biologist and Nobel laureate George Wald had no doubts. He said: " I tell my students, learn your biochemistry here and you will be able to pass examinations on Arcturus." Harold Morowitz, a biologist at George Mason University near Washington, D.C., points to the fact that "there are only four different kinds of one-carbon compounds." That severely limits the number of ways of building up and breaking down larger molecules. Others, like Christopher Chyba of the SETI Institute, urge caution in drawing too many conclusions about the small print of biology elsewhere. Again, it's the problem of one data point.

Details aside, astrobiologists agree that the most promising places to look for life will be those where carbon-based molecules have had a chance to collect and become concentrated. Two other ingredients have also been singled out, more or less unanimously, as key biological prerequisites: the availability of liquid water and a suitable energy source that can be tapped by the metabolism of living things. Intriguingly, all of these commodities—organic matter, water, and metabolically useful energy sources—are starting to look pretty common in the universe. But whether life proves to be plentiful or not beyond the Earth depends crucially, too, on a number of other factors. Most important, there's the question of abiogenesis. How easily, given the right raw ingredients, does life arise?

2
Original Thoughts

Four billion years ago, it was Hell on Earth—or so many scientists believe. Direct evidence is hard to come by, because our world conceals its distant past well. The oldest known outcrop of terrestrial rock, the Acasta gneiss 350 kilometers north of Yellowknife in Canada's Northwest Territories, dates back slightly less than four billion years, leaving the first half billion years of Earth's history without a rock record at all.* To make matters worse, even those fragments of ancient crust that remain, like the Acasta gneiss, have been so melted and distorted over the ages that they carry only a cryptic record of their infancy.

Fortunately, to get some idea of what the young Earth might have been like, geologists aren't confined to studying the ground beneath their feet. Other clues come from space—from our neighboring worlds, in particular the Moon.

Untouched by weathering or the relentless shifting of continents, the lunar surface bears witness to a ferocious battering it took during the very time the geological record is missing or hard to read on Earth. Those dark, round patches that suggest a human face are in reality wide basins, now lava-filled, that were excavated by colliding asteroids many tens of kilometers across. Elsewhere in the solar system, every other planet and moon, including the Earth, is thought to have been bombarded during this holocaust age by debris left over from the initial bout of world-making.

It was the most inhospitable time imaginable, fittingly christened the *Hadean* or "hellish" era by American paleontologist Preston Cloud. A number of big asteroids (no one knows how many) are reckoned to have slammed

---

Although, as we'll see, some microscopic samples, such as zircon crystals, are known with ages greater than 4 billion years.

into the Earth at tens of thousands of kilometers per hour, gouging out craters as wide as medium-sized countries. Each ensuing blast would have ripped away part of the atmosphere and replaced it with a searing swathe of vaporized rock hot enough to turn ocean water to steam. The very idea of life beginning or surviving under such circumstances seems absurd. And yet, against all the odds, powerful evidence has come to light that the Hadean era was *not* sterile.

What actually went on in that strange, long-ago time? And how, in the midst of it, did life on Earth come about? Under what conditions, by what means? The answers are important to us personally because they are part of our story, our heritage. But they are crucial too in a wider context. What happened here, on this planet four billion years ago, was not a miracle or a fluke. It was a particular instance of how matter and energy can spontaneously engage in a runaway process of spiraling complexity, until chemistry shades into biochemistry, and biochemistry crosses the threshold into life. What triggers that snowball effect? What combination of factors trips the life-generating cascade? The search for our earliest roots is an essential part of a larger quest that embraces the universe as a whole. That is why the science of the origin of life on Earth lies at the heart of astrobiology.

✵

In a letter to the botanist Joseph Hooker in 1871, Charles Darwin wrote, "[I]f . . . we could conceive in some warm little pond, with all sorts of ammonia and phosphoric salts, light, heat, electricity, etc., present that a protein compound was chemically formed, ready to undergo still more complex changes."

For a mid-Victorian raised on rigid creationism, it was a bold, insightful idea—that life could arise naturally, by stages, from lifeless chemistry. Darwin was modest enough, and aware enough of the technical limitations of his time, not to take his conjecture too seriously. But the notion of a "warm little pond" would come increasingly to dominate scientific thinking on life's origins.

After all, what alternative explanation could there be? Water, the right blend of "salts," a source of energy to drive the synthesis—these were the essential ingredients as seen by Darwin and by every scientist today. Given that basic triad, where on the young Earth could life have first emerged, but in some sun-drenched, chemically-steeped body of water?

Of course, the concept needed fleshing out. The chemical stepping stones leading to life had to be firmly laid down—and ultimately traversed in the lab; otherwise, it was no more than a picturesque surmise. Three ideas in the first half of the twentieth century gave Darwin's pond the scientific credibility it needed. They were ideas about the nature of the infant Earth. That was crucial, because talking sensibly about how life began can only start when the conditions are known. What was happening on the surface? What was the atmosphere like?

First, the atmosphere was *reducing*; in other words, it was rich in hydrogen and other hydrogen-containing gases, such as methane and ammonia. Reducing chemicals are important because they donate electrons to other substances and thereby produce energized molecules. These molecules are then able to take part in chemical reactions that can lead to the creation of more complex substances. In 1924, the Russian biochemist Alexander Oparin first linked the ideas of a reducing environment with the chemical origin of life in an obscure little pamphlet published in Moscow. Astronomers had found reducing atmospheres on Jupiter and the other giant planets, and these worlds were thought at that time to be in an earlier stage of development than our own. In a sense, they were believed to offer a window on the Earth's past. Oparin suggested that reducing chemicals in the Earth's primordial atmosphere had found their way into the ocean waters below, where they reacted and combined to form substances of increasing complexity. And so, in time, simple life emerged.

Four years later, and quite independently (Oparin's work having not yet reached the West), J. B. S. Haldane speculated along similar lines. But Haldane added the second factor into the nurturing milieu for life—ultraviolet radiation. There would have been no free oxygen in those early days, because oxygen comes from plants and other light-harvesting creatures. Without oxygen or the protective ozone layer that forms from it, the Sun's harsh ultraviolet radiation would have beaten down relentlessly on the surface. Lethal today (even mild UV can give us sunburn and skin cancer) high-energy sunlight became Haldane's spark to animate the oceanic "hot, dilute soup."

But the diluteness was a problem. If the chemical building blocks were spread over the whole ocean, what were the chances of them combining to make something so elaborate as life? The soup had to be condensed. In 1947, Irish physicist J. D. Bernal suggested how. In a classic paper, called "The Physical Basis of Life" and delivered before the British Physical Society in

London, Bernal argued that lagoons and pools at the margins of sea and land served to concentrate the prebiotic brew. Clay deposits, too, perhaps played a part, providing a surface on which molecules could congregate and so more easily interact. But Bernal's vision extended far beyond the confines of Earth. In the same lecture he declared that, "[T]errestrial limitations obviously beg the question of whether there is any more generalized activity that we can call life . . . Whether there are some general characteristics which would apply not only to life on this planet with its very special set of conditions, but to life of any kind."

To Bernal, such cosmic thinking came easily. In 1929, in *The World, the Flesh, and the Devil*—a little book dense with extraordinary, brilliantly original, often wild ideas—he examined subjects ranging from space colonies to starships, from non-corporeal life-forms to the future of intelligence in the universe. Parts of it sound futuristic even now. In any event, a couple of decades later, elevating the discussion of life and its origins to the universe as a whole was no daunting step for him. In 1952, he put the case even more plainly in a speech to the British Interplanetary Society: "The biology of the future would not be confined to our own planet, but would take on the character of cosmobiology."

So now the three theoretical refinements needed to breathe life into Darwin's pond are on the table: a reducing atmosphere, a sharp energy source (ultraviolet radiation and possibly lightning), a means of concentration. Oparin, Haldane, and Bernal have had their say, and Bernal is talking about the globalization of biology. What is happening elsewhere? In the United States and in the Soviet Union, the Space Age is rushing toward the launch pad, suddenly making the science of life on other worlds seem a practical proposition. In Alma-Ata, Kazakhstan, the astrophysicist Gavriil Tikhov, who arrived here with many other evacuee academics during the Second World War and remained, is looking for evidence of Martian vegetation. In 1953, he publishes a book that few Westerners will ever read, but whose title is a word used for the first time: "Astrobiology." Half a world away, in Chicago in that very same year, the young Stanley Miller seals Oparin's reducing atmosphere in a flask, throws a switch, and puts Darwin's conjecture to the test.

Miller's advisor at the University of Chicago, the Nobel Prize-winning chemist Harold Urey, had tried to talk his Ph.D. student out of doing the experiment. Not that there seemed anything wrong with it in principle; Miller was simply following the ideas about the origin of life that he'd heard

Urey lecture on a couple of years earlier. Urey just thought that trying to recreate the "warm little pond" in the lab was too risky and time-consuming for a doctoral project. But Miller won his grudging permission to try; they'd give it six months, maybe a year.

Within a *few weeks*, Miller had his apparatus and his first remarkable results. To a flask containing a gassy mixture of methane, ammonia, hydrogen, and water vapor—Oparin's dawn atmosphere—he injected 60,000-volt sparks to simulate lightning. The byproducts were allowed to condense and were collected in a glass U-bend that simulated a body of water on the Earth's surface. Heat supplied to a second flask connected to the U-bend recycled the water vapor, just as water evaporates from lakes and seas before moving into the atmosphere and condensing again as rain.

Who'd have given odds it would work? And yet, on the time scale of Biblical creation, Miller's genesis-in-a-bottle spawned some of the raw materials of life, including amino acids—the chemical sub-units of proteins.

Miller presented his results at a crowded seminar. Urey sat in the front row alongside the great physicist Enrico Fermi. During the question period that followed, Fermi turned to Urey and said, "I understand that you and Miller have demonstrated that this is one path by which life might have originated. Harold, do you think it was *the* way?" To which Urey replied, "Let me put it this way, Enrico. If God didn't do it this way, he overlooked a good bet!"

God, however, may have spread His wager. Again, we need to ask: What conditions existed on the young Earth? What was the environment like at the time that life is supposed to have started? Until we know that, everything else is guesswork.

In 1953, when Miller and Urey shook the scientific world, the age of the Earth was uncertain. Estimates ranged from about three and a half to four billion years. Today, it's put pretty reliably at 4.55 billion years. In 1953, the most ancient fossils known were less than two billion years old. Today, there's evidence for life stretching back almost twice as far.

But the crux of the matter is this. The Miller–Urey experiment, and any conclusions that follow from it, stand or fall on the strength of one claim—that the Earth had a reducing atmosphere. At the time the experiment was carried out, that was a reasonable assumption. Not everyone agreed with it, but it was easily defensible. The experiment isn't even fussy about what reducing ingredients are put in the sparking flask. You can get the same biologically promising brown goop that Miller obtained if you use, say, hydrogen, ammonia, and carbon dioxide as the starting mixture. The synthesis won't be quite as efficient, but it will work. On the other hand, *it will fail completely*—there'll be no amino acids and other organic goodies produced—if the gases in the sparking flask are not reducing at all; that is, if hydrogen, methane and ammonia are all absent.

Miller, though semi-retired, remains involved in his lab, only today he is at the University of California in San Diego. Almost fifty years after his miracle in Chicago, he and quite a number of other scientists around the world are still working to show how the chemicals needed for life can be built up in some version of Darwin's warm little pond. Call them the "surface, sunlight" guys. Their chemical syntheses are every bit as far along as those of any other researchers working on the origin of life—which is to say, not very. But that isn't the issue. The issue is that every one of their results hinges on the question of the reducing atmosphere. And what is the consensus among geologists and geochemists? That the atmosphere on the young Earth was gassed out of the planet. In other words, its composition was very similar to the mixture that belches out of any volcano today: mostly carbon dioxide, carbon monoxide, nitrogen and water vapor with a smelly dash of sulfur compounds. Such an atmosphere will rob energy from other compounds and interfere with reactions that transform simple organic substances into more elaborate ones.

Where does that leave the "surface, sunlight" guys? It leaves them without a bowl of prebiotic soup. Without the reducing gases, no matter how clever their choice of pond or surface energy source, they can't get the chemical build-up off the ground. The only way a surface, sunlight guy can save the day, if deprived of a reducing atmosphere, is to accept a hand-out of simple organic material—amino acids and so on—from somewhere else. Once this food parcel of basic organics is dropped in, it's possible to turn on the solar UV lamp and choose some beach or other promising spot and push on with making more complicated things, like short strands of protein or the building blocks of DNA. But for those of the unreformed surface, sunlight school, a proven absence of reducing gases would suggest a change of career.

☀

Still, we know life came from somewhere. If we've had a good, liberal education, we know that life evolved. It must have had its origin at some point. But where, when, and how?

It's very natural to suppose that life started out on the surface, in the sunlight, at some moderate temperature, because that's what is friendly and clement and nurturing to the kind of life *we* are. Common sense suggests that the last place to look for life and its origins would be deep beneath the surface, in the dark, at outrageously unsociable pressures and temperatures. But a couple of decades ago, common sense went out the window.

Geologists had guessed there might be hot gushing springs on the ocean floor near where molten rock pushes up through the crust. Their suspicions were confirmed in 1977 when the first deep-sea vents were sighted in the search-lights of the submersible *Alvin*, about 2,600 meters down off the Galapagos Islands. But what researchers hadn't bargained for was that there might be life around these ocean-bed geysers. In fact, they found not just life but an amazing zoo of creatures swarming in the environs of the vents. Here, where no sunlight ever reached and the pressure was immense, were bizarre, nine-foot-long tube worms without mouth, gut or anus, as well as unusual crabs, clams, and mussels, most of them white—hundreds of species in all, 95 percent of them new to science.

At the bottom of the food chain, as dense as a blizzard in the water and as thick as a carpet on the walls of the vents and the nearby sea floor, were microbes. Not ordinary microbes, of course—not in a place like this. Many of them were thermophiles (heat-lovers) or hyperthermophiles (extreme heat-lovers), living right next to the stream of scalding water, at temperatures of 100°C or more. No need to look far for their source of energy and raw materials. Here it was, right in the devil's brew of chemicals gushing out of the vents—hydrogen sulfide (the gas that smells of rotten eggs), metal-rich compounds, and carbon dioxide dissolved in their hot water surrounds.

The first person to look upon this astonishing menagerie was the oceanographer Jack Corliss, aboard *Alvin* on its historic discovery dive. If life could thrive in such a place, he reasoned, why shouldn't it also have started here? There was energy in plenty, water, a concentrated source of chemicals including carbon (from the dissolved carbon dioxide)—the basic ingredients for getting biology underway. Hydrothermal vents would almost certainly have existed on Earth at a very early stage, as soon as the planet's outer crust had hardened and been covered by ocean—some time between 4.3 and 4.4 billion years ago. It all made sense, and in that dawning realization, Corliss turned his back on Darwin's little pond. He became, instead, a "deep, dark" guy.

What Corliss and others started to grasp in the 1970s is an entirely different way to think about the origin of life. It can't be called the right one, because no one knows enough to make that judgment yet. But it's a reasonable alternative in a field that the surface, sunlight guys had had to themselves for a long time. Suppose the first organisms *didn't* depend on the Sun and atmosphere—the things above—but on geothermal energy and nutrients welling up from below.

The discovery of deep-sea vents was just one of the triggers of this strange new paradigm. Heat-loving bacteria had already been found in the sulfurous hot springs of Yellowstone National Park in Wyoming. After that breakthrough, it seemed that wherever researchers looked—in rocks deep underground, in very salty, caustic or acidic pools, in all sorts of crazy places hostile to ordinary life—they discovered different types of hardy microbes. A huge portion of the Earth's livestock had been overlooked, because no one dreamed such creatures could exist. Yet here were these "extremophiles," not only surviving in their exotic little realms but in many cases incapable of surviving anywhere else. And the majority of them lived off energy and inorganic materials coming from *inside* the planet.

Well, so what? These eccentric organisms could very well have evolved from creatures that once lived under more normal circumstances on the surface. If we're asking what came first, then we need some reliable way of working out the details of the family tree of life. What are the oldest organisms on Earth? What was our ultimate ancestor like?

In the late 1960s, biologist Carl Woese at the University of Illinois began retracing the genealogy of life using a new technique that he'd devised. It involved looking at certain genes that hold the instructions for making a molecule that's found in the cells of every living thing: small subunit ribosomal RNA, or to use its snappy acronym, SSU rRNA. Woese singled out this particular molecule because it lies at the core of the machinery inside the cell where proteins are put together. Since proteins are fundamental to every kind of life we know, it's reasonable to suppose that SSU rRNA has been around as long as life itself.

Woese realized that if he compared the genes coding for SSU rRNA in many different species, they would provide a measure of how far apart different species were in evolutionary terms. Given enough data, the structure of the tree of life would emerge in unprecedented detail. Collecting these data required many years of tedious work, but by the 1980s Woese and others working on the problem had learned two things—and it's hard to say which is more surprising. The first is that the main branches of the tree of life need to be redrawn. Instead of five or six kingdoms as biologists had previously thought, the genetic evidence puts all life-forms into three bigger branches or *domains*: bacteria, archaea (a separate category of microbe), and eukaryotes (including animals, fungi and plants). The second surprise is that the deepest, most ancient part of all three branches, near where they first grow apart, contains only one type of organism—the hyperthermophiles found in boiling-water springs and deep-sea vents.

The obvious conclusion is that the original ancestor of every living thing on Earth was a hyperthermophile, but that may be going too far. Although hyperthermophiles seem simple by human standards, they're actually incredibly complex. *Every* living organism is incredibly complex compared with whatever the first life-forms on this planet were.

Even so, some scientists now argue that the genetic evidence favors a high-temperature origin. Others, especially of the surface, sunlight school, reply that it's impossible to draw any conclusions because the kind of hyper-thermophiles we know about are the products of a long line of evolution, about which the genetic data say nothing. Maybe the hyperthermophiles are the survivors of an era in which primitive microbes lived in all sorts of different environments. Depending on where your loyalties lie, score that advantage to the deep, dark guys, or keep it at deuce.

Time to serve up the fossil evidence. What do the actual remains of creatures in ancient rocks say about the origins of life? Remember, we're not dealing here with *T. R*ex skulls or even delicate imprints of fish scales. The earliest traces of life are incredibly hard to find, even if you're looking in the right place. Often they're so battered, blackened, and generally past their sell-by date that it's hard to tell if they're even biological. Still, enough solid biogenic material has been unearthed to add tantalizingly to the origins debate.

June 2000 brought news that Birger Rasmussen, a paleontologist at the University of Western Australia, had discovered fossils of single-celled organisms in rock 3.2 billion years old. The tiny filaments, a thousandth of a millimeter in diameter and a tenth of a millimeter long, turned up in core samples drilled in the Pilbara region of the northwestern Outback. Nothing too amazing about that—older fossils have been recovered from ancient sediments elsewhere. What's unusual is that Rasmussen's filaments were found in igneous rather than sedimentary rock. Based on their location and appearance, which are similar to those of microbes found around deep-sea vents today, Rasmussen concluded that they'd probably lived in the pores and crevices of rocks at shallow depths below the sea floor. Advantage deep, dark guys?

Not so fast. There may have been life around hydrothermal systems more than three billion years ago, but there were also microbes living off sunlight in shallow waters *three and a half* billion years ago. Their remains were dug up in 1993, also in northwestern Australia, in the Warrawoona region, by UCLA paleobiologist William Schopf. Deuce.

That the Warrawoona fossils are of light-catching bacteria is particularly interesting. Photosynthesis is a very sophisticated means of trapping energy that would have been completely beyond the earliest organisms on Earth. Whatever camp they belong to, biologists stand together in the belief that the first life-forms got their energy from chemicals, either organic (as in a prebiotic soup) or inorganic. They definitely weren't light-catchers. No one knows how long it took photosynthesis to evolve once life had appeared, but a few hundred million years is considered reasonable. Using this figure as a guide, the Warrawoona fossils suggest that life couldn't have got going much later than about four billion years ago. That pushes us back into the Hadean era, the supposedly hellish time when the Earth was a punching-bag for wayward comets and asteroids flying through the inner solar system. Those were tough times for sun-worshippers. But they were perhaps survivable for vent microbes or subterranean rock-dwellers. A thick layer of ocean or crust would soften the effects of city-sized asteroids falling on the surface. Advantage deep darkers?

Maybe, but we're not done yet with what very old rocks can tell us. The Warrawoona finds take us to the present known limit of *morphofossils*—fossils in which some shape or structure of the original organism can be made out. But the possibility of *chemofossils*—the smashed-up chemical remains of living things—offers a portal on even earlier times.

From Australia, the story moves to Greenland. With the exception of the Acasta gneiss in Canada, the oldest known rocks on Earth occur in the southern part of West Greenland, not far from the capital city of Nuuk. Some of them outcrop on the little island of Akilia, others as the so-called Isua formation on the mainland. During a 1991 expedition, geologists Clark Friend, of Oxford Brookes University in England and Allen Nutman of the Australian National University in Canberra collected a mineral sample from Akilia, which Nutman later dated at a spectacular 3.87 billion years. That would make it the oldest water-lain sediment ever found. But could rock that had undergone so much change over the eons—even rock that had started out sedimentary—possibly harbor traces of life?

At the Scripps Institution in California, oceanographer Gustaf Arrhenius and his research student at the time, Stephen Mojzsis, had been looking for very early signatures of life on Earth using a method that involved examining microscopic grains of graphite, a variety of carbon, enclosed within crystals of the mineral apatite. The apatite serves as a kind of miniature vault, protecting the graphite from the ravages of time. Both graphite and apatite *can* be formed either inorganically or by living things. A way to tell the difference is to look at the ratio of the two isotopes of carbon that the graphite contains. Carbon comes in three forms, including carbon-12 and the less common carbon-13.* Some life processes—some aspects of metabolism—favor carbon-12, so the remains of once living organisms tend to have a slighter higher ratio of carbon-12 to carbon-13 than occurs in the non-living world. Just such an elevated ratio was found by the Scripps researchers when they analyzed a sample of graphite from the Akilia rock. A second sample, from the Isua formation and dated at 3.75 billion years, also tested positive. They're the oldest signs of terrestrial life ever found.

There are a couple of caveats. Some scientists have cast doubt on the heightened carbon-12 levels, suggesting they might have an inorganic origin. The other point is that the age of the Akilia rock, 3.87 billion years, is currently disputed. In reply, Stephen Mojzsis says:

> There is confusion about the ages of the rocks on Akilia island that derives from work carried out on the wrong samples and published [in 1999] in the journal *Chemical Geology*. Subsequent papers . . .by V. R. McGregor and A. P. Nutman et al . . . have resolved the issue, and the rocks are most likely >3.84 billion years in age. As for the biogenicity of the carbon isotopic signature, much of the debate there is generated by those who know almost nothing about isotope geochemistry. Let me emphasize . . . that there exists no *known* abiotic/geological process that can mimic the large isotopic fractionations of carbon by life. I am unaware of any discussions published in peer-reviewed journals (rather than whispered in hallways, or in someone's book like "Cradle of Life" by Schopf) that seriously proposes an abiotic origin of carbon isotopic excursions up to 6% in the geological record.

In any event, the 3.75 billion-year age of the Isua sample is uncontested. This is still a couple of hundred million years older than the previous record-holder, the Warrawoona fossils, and suggests that quite sophisticated life, comparable to some modern microbes, was already well established just 800

---

*A third isotope, carbon-14, is used to date archaeological remains, but decays too quickly to be useful at the time-scales we're interested in.

million years after the Earth formed. That would mean life originated during the supposed period of intense cosmic bombardment.

There's also this to consider. The ancient Greenland rocks we've just been talking about were found alongside what are called banded iron formations (BIFs). These BIFs are thought to have built up as layers of sediment *around hydrothermal sources*, where iron- and other metal-rich compounds are commonly deposited. Game, set and match to the deep, dark gang?

Some researchers are leaning that way. If you've got traces of early life in a geothermal setting in the midst of a cosmic Blitz, what are you supposed to believe? Who's survival chances do you rate better: a sitting duck in a pond on the surface, or a clever duck sheltered by several thousand meters of ocean water or rock?

But before we crack open the case of champagne, we'd do well to pause and listen. In the background, behind the cheering crowds and press photographers, a few voices are asking, quietly but insistently: *What bombardment? What makes you say it was Hell on Earth, four billion years ago?*

The question comes as a bit of shock. Wasn't the early solar system like planetary bumper cars? Could the Moon have gotten so many giant impact basins and craters while, next door, the Earth escaped? If anything, the Earth ought to have been pummeled even more because its stronger gravity would have accelerated incoming objects to higher speeds. It all seems reasonable enough. The point the dissenters are making though is: Where's the evidence? Where *on Earth* is there a clear sign of heavy bombardment?

When you look at the impact record on the Moon, it builds up to a peak between 4.0 and 3.8 billion years ago and then tails off until 3.45 billion years ago. The banded iron formation associated with the Isua rock is around 3.75 billion years old. As the oldest undisputed sedimentary record we've got, it should contain some trace of the turmoil that was supposedly going on at this time. But its testimony is inconclusive. There's some evidence of grading—successive layers of the Isua BIF containing different sized particles—that could be due to sorting and settling of rock fragments after a violent event such as an impact. However, an impact-related deposit ought also to be accompanied by an enrichment of elements, such as iridium, that are characteristic of asteroidal material. No such enrichment in these ancient rocks has yet been observed.

So we need to keep an open mind. All the excited speculation in recent years about the Earth having been pounded by giant asteroids during the time that life was starting out isn't as securely based as it seemed. The devastation that happened on the Moon might have been localized.

Let's be clear about this. No one is saying that there weren't a lot of collisions in the early days of the solar system. You only have to look at the faces of other worlds—Mercury, Mars (although there's been much erosion here), and the moons of the outer planets—to see how pockmarked they are. And most of this cratering is very old. The solar system was definitely a hard-hat zone during its first billion years. What's at issue is not the general bombardment, but the occurrence and frequency of really big impacts like those that carved out the lunar basins. Was the so-called *late heavy bombardment*, which the Moon undoubtedly suffered, purely a lunar phenomenon or was it widespread? Did it affect the Earth?

The weight of scientific opinion is that the Earth was hit at least once very hard indeed—in the collision that created the Moon. According to the most popular version of this theory, about 4.5 billion years ago the young Earth was side-swiped by another planetary object, perhaps bigger than Mars, that splashed out a great fountain of terrestrial matter. Much of this lost material settled in orbit and eventually pulled itself together to form the Moon. There's also a rival theory, out of favor at the moment, that says the Moon was once an independent body that was captured when it strayed too close to the Earth. But for the sake of argument and drama, let's stick with the cataclysm scenario. Under what circumstances could the Moon have suffered its late heavy bombardment and the Earth got away more or less unscathed?

It could have happened if the Earth already had a few smaller satellites— moonlets—before *the* Moon came into existence. Once the Moon had formed in orbit around the Earth, it would have started to slow the Earth's rotation due to "tidal braking." At the same time, in order to conserve the total angular momentum of the system, the Moon would have gradually accelerated, causing its orbit to enlarge. As the Moon sped up and drifted out (by about three centimeters per year), it would have caught up with and smashed into any moonlets that happened to be around. According to some calculations, the most likely locations for ancient Earth moonlets were roughly at the distances the Moon would have moved out to at the time of the late bombardment, which would explain the big impact basins.

Ideas like this sound a note of caution. We don't have to buy into the notion (however viscerally exciting) that the Earth was struck a number of times during the Hadean era by giant asteroids, perhaps violently enough to sterilize any surface life. True, it must have been pounded, often and sometimes pretty hard. But it wasn't necessarily struck repeatedly by asteroids measuring, say, hundreds of kilometers across. Purely from the point of view of impacts, we can't rule out the possibility of life—even surface life—at the four-billion-year mark or earlier.

We also can't rule out the possibility that life arose a number of times and on each occasion (except the last) was wiped out. The rock record of the Earth's first billion years is so fragmentary that several giant, sterilizing impacts *could* have occurred, and we'd have no way of knowing about them.

We can't even be sure how the Earth formed, and therefore what it was like during its first few hundred million years. What's pretty well beyond doubt is that the Earth, like all the other planets, grew out of a spinning pancake of dust and gas that surrounded the young Sun. The little particles of dust in this cloud bumped into one another and clung together by chemical stickiness. Later, when the clumps of dust became big enough—somewhere between the size of large boulders and small mountains—their self-gravity became strong enough to pull in more material much faster. But the details of that accretion process are still poorly understood.

The "standard" model of how the Earth formed depicts our planet at the start of its career as a brutally hostile place. Its surface was a roiling magma ocean kept molten below by heat pouring out of the interior due to the decay of radioactive elements and the release of gravitational energy, in addition to the pounding it was getting from space. Not until the surface had cooled to the point where liquid water could collect without being boiled away was the scene even remotely set for life. Given the Isua evidence of complex life by 3.75 billion years ago, that puts an awful lot of time pressure on any theory of life's origins. A different model for the Earth's formation, however, allows for more leisurely biological development. According to this alternative theory, the accretion took place more slowly: the iron core building up first, followed by overlying layers of rock, so that heat inside the planet had more time to escape. In this scenario, the Earth never had a molten surface and therefore would have been inhabitable much earlier on.

Could life have started well before 4 billion years ago? Perhaps so, according to a discovery announced in January 2001. A team of scientists from Scotland, the United States, and Australia reported having found a tiny crystal of zircon (a substance containing silicon, oxygen, and zirconium, among other elements) in northwestern Australia dated at 4.4 billion years old. Analysis of the oxygen isotopes suggested that the crystal could only have originated in a wet, low-temperature environment. If the Earth really was a water-world so long ago, then the possibility of extraordinarily early biological developments needs to be taken seriously.

✷

So many theories, so little data! But that's always the case with a new science, and astrobiology is no exception. What we have to ask is: at this stage, where is the scientific consensus about the origin of life? Given the evidence in hand, what do researchers think is the most reasonable theory about the time, place and means by which life first emerged on Earth?

It's becoming hard to avoid the conclusion that hydrothermal systems played a significant role. Circumstantial though it may be, the genetic evidence leans toward early heat-loving life-forms, and the geological evidence leans toward early life in a hydrothermal environment. Deep ocean vents are particularly in vogue right now, partly because they're so wonderfully weird and evocative as potential founts of life. But it isn't just that they make good copy; as potential genesis machines, they have created a huge amount of interest in the international research community. Much of the credit goes to the innovative ideas of one man: Günther Wächtershäuser (pronounced "vecter-shoyzer").

Wächtershäuser is a German scientist (an organic chemist at the University of Regensburg) who devised a theory that some believe will revolutionize origin-of-life research. In a disarmingly simple approach, he asked: What is most basic about living things? Whatever is most basic surely had to be in place right at the start. And the answer he gave seems obvious when you think about it, though it took a spark of genius to see it first. Before creatures can make copies of themselves or grow, what must they be able to do? *Metabolize*. Metabolism, argued Wächtershäuser, is the key to the origin of life, because it must have preceded everything else, even replication.

Here, then, was a new paradigm to set alongside that of the deep, dark school. And in this case, "alongside" can be taken literally. Having asked what is most basic about life, Wächtershäuser went on to ask what is most basic about metabolism. When you look at the great metabolic map of life, what lies right at its heart? For an organic chemist there can be only one answer: a little cycle of reactions that runs like a watch spring inside every cell of every organism—the citric acid cycle (or, as it's also known, the Krebs cycle).

This loop of about nine chemical reactions, the same in all creatures on Earth, generates an instantly accessible, internal energy supply from simple organic compounds. Now, if instead of maintaining life you want to *make* it, there has to be an adjustment. Instead of running the cycle forward, in the normal way, you have to run it *in reverse*. In other words, you need to pump energy into the little roundabout of reactions so that it becomes a manufacturing plant for basic organic materials. Then you have the building blocks from which to move to the next stage of assembly.

That was the first of Wächtershäuser's clever ideas. The second was to realize where it might work best: alongside hydrothermal vents spouting just the right kind of energy-rich compounds to fuel the backward-running citric acid factory.

Today, these ideas and others like them are being put to the test in the lab, with encouraging early results. At the Carnegie Institution in Washington, D.C., inside something called the "bomb," built using steel panels from a scrapped battleship, mixtures of chemicals are subjected to the kind of ferocious temperatures and pressures found in a deep-sea vent. In this way, Carnegie researchers have managed to make ammonia, a key ingredient in Wächtershäuser's scheme, under simulated deep-sea vent conditions, and then used ammonia to leap-frog to the amino acid alanine. At about the same time, early in 1999, Koichiro Matsuno and his team at Nagaoka University announced results from the world's first artificial deep-sea vent. Fed with the amino acid glycine, it churned out connected strings of amino acids—short chains called oligopeptides—that are the next milestone along the road to proteins.

There's still a long way to go. If synthesizing life in the laboratory is an Everest expedition, scientists are roughly at the stage of boarding the plane to Kathmandu. They've shown how *some* of the small molecular sub-units of life can be made under conditions that might have existed on the young Earth. Ahead now lies the greatest challenge: to fashion, from the bottom up, proteins and nucleic acids, with their intricate structures and many thousands of component atoms—again under feasible primordial conditions.

Something along the lines of Wächtershäuser's hydrothermal assembly plant is seen by some researchers as offering the best way forward in the stages of synthesis to come. But there's been great progress, accompanied by a feeling of imminent breakthrough, on several fronts. Remember that hydrothermal systems exist at ground level too. And if the Earth escaped really devastating impacts, there's no reason to suppose life couldn't have survived on the surface at the interface between the geothermal and solar regimes—in runoff pools, in bubbling springs, and so on. In fact, even if the Earth was completely sterilized, perhaps several times over, there's nothing to say that life didn't reemerge Phoenix-like from the ashes of its former self. Very little can be ruled out at this stage. The slow accretion model of the

Earth even allows for the possibility of reducing gases building up in the early atmosphere, so that Miller-type production of basic biochemicals comes back into the picture. The surface, sunlight guys are still very much in business.

Nor do the possibilities end there. At one extreme of the environmental spectrum, researchers such as Cornell's Thomas Gold continue to argue the case for life's origins far underground inside microscopic pores of rock at high temperature and pressure—the "deep, hot biosphere" milieu. At the other extreme, intriguing new evidence has come to light that life may have begun in the air. Aerosol particles from ocean spray, lofted high into the atmosphere, have been found to contain a concentrated organic mixture sealed within a thin skin of fatty molecules. Announcing this discovery in July 2000, an international team of scientists argued that such droplets, which resemble primitive cells, could provide ideal reaction vessels for biochemical synthesis.

The origin of life field is entering something like a postmodern era. Far from narrowing the number of places where life might have started out, scientists are beginning to see that there are many different viable possibilities. There is a growing suspicion that life doesn't need much encouragement—some water, an injection of energy, a huddling of carbon chemicals, and voila! If the trend evident today continues, we may find there are many roads to life. Underground, undersea, in surface waters, in pools and lagoons, on the moist surface of minerals, in the air—perhaps the build-up to life can and does occur, even simultaneously, in all these environments. The very speed with which life appeared on Earth suggests it grasps at the slightest chance to take hold.

But that speed of development also suggests the origin process may have been kickstarted in some way. And there are other reasons to suspect the intervention of an outside agent. A few of the basic building blocks of life have proven stubbornly difficult to make in the lab under early Earth-like conditions. What's more, a curious feature of the molecules at the heart of all terrestrial life has eluded explanation by any of the theories we've looked at so far.

Four billion years ago, it may have been Hell on Earth. But the same intruders from space that were such a hazard to anything in their path may have been among life's greatest benefactors. It's time to consider our cosmic connection.

3

Star Seed

Every day, unheralded and largely unseen, hundreds of tons of dust and rock land on Earth from space. Most of this alien rain is either too small to notice or comes down where nobody is around. But occasionally, extraterrestrial debris makes a big impression.

Around 50,000 years ago, a body of iron and nickel weighing millions of tons slammed into the dry plain near present-day Winslow, Arizona, and carved out a rocky amphitheater almost a mile wide and as deep as a 60-story building. In 1908, what seems to have been a chunk of a comet exploded over Siberia, felling or stripping hundreds of square kilometers of trees, burning reindeer to death, and sending the tents of nomads flying through the air. Had it fallen over a major city, the results would have been catastrophic.

More recently, on January 18, 2000, a meteorite the size of a small truck and weighing at least 200 tons streaked across the skies of northern Canada, broke up in the atmosphere, and scattered thousands of fragments near Lake Tagish in the Yukon. Scientists were delighted when pieces of this object were recovered in a still-fresh, frozen state, because the Lake Tagish meteorite is of a rare and important kind. Locked within it, billions of years old, are carbon chemicals that formed in the lonely void between the stars.

Today the amount of material delivered from space each year is quite small. But in the remote past, the rain of extraterrestrial debris would have been a torrent. What role did it play in biological developments here?

✵

As early as the first half of the nineteenth century, scientists knew that some meteorites contained organic matter. Called carbonaceous chondrites, they account for only a small percentage of all meteorites found, yet they offer vital clues in the search for origins of worlds and of life. The trouble is, unless a meteorite is recovered as soon as it's landed, there's the likelihood of contamination. Then it becomes fiendishly hard to tell imported organics from the mundane variety. In recent times, though, scientists have become much better organized in their efforts to track down and capture pristine samples that are newly arrived from space. And they've been helped by a couple of lucky falls near centers of population.

One of their biggest coups, before the Yukon encounter, was the quick recovery of fragments of a carbonaceous chondrite that exploded over the town of Murchison, 400 kilometers north of Perth, Australia, in September 1969. Over 80 kilograms of the Murchison meteorite were found and taken into custody. And indeed, what followed was something like a major criminal investigation. Painstaking analysis by teams of investigators over three decades uncovered some surprising fingerprints—clues that led ultimately to a possible solution to one of life's long-standing mysteries.

The molecules of many organic substances, including amino acids and sugars, come, like gloves, in mirror-image forms known as enantiomers. In most of nature, chemicals with this property occur in racemic mixtures, meaning that the numbers of left- and right-handed molecules are equal. But the cells of living organisms on Earth maintain a curious and absolutely rigid prejudice. *Every* amino acid is left-handed* and *every* sugar is right-handed. It isn't the sort of fact you can ignore for long. Nagging questions intrude: What caused such a strange bias? What might it tell us about the way biochemistry came about on our world? Was the chemical deck stacked before life began, or did the trend to one-handedness emerge during the early stages of evolution?

None of the lab experiments ever done to mimic the prebiological world moves us closer to an answer. Any amino acids they've produced have always, without fail, been split evenly between right- and left-handed forms. Something's been overlooked; some piece of the puzzle of life is missing.

Researchers had reported finding amino acids in meteorites as early as the 1950s but the suspicion of contamination—in some cases vindicated—always loomed over the claims. Not for another couple of decades could scientists bring themselves to accept the reality of amino acids from space. The Murchison meteorite played a prominent role in convincing them. Inside it

---

*With the exception of glycine, the simplest amino acid, which doesn't show handedness.

and a few other carbonaceous chondrites were found several dozen amino acids that had never been seen before in nature. Where else could they have come from but beyond Earth? Furthermore, all the Murchison amino acids, *including those that matched terrestrial varieties*, seemed to be in racemic mixtures, ruling out the possibility they might be Earthly contaminants. A space origin became unavoidable.

What did this new discovery imply? First of all, that surprisingly complicated molecules can form under conditions very different than those on a planet's surface. Evidently, there is an assembly process "out there" that is much more sophisticated than most people had suspected. What's more, it is capable of mass-producing some of the substances—amino acids—that are the bedrock of the kind of life we know. At the very least, this meant that certain biochemical raw materials are a common cosmic commodity. And it could be taken further: It could be taken as a sign that wherever life appears, it will likely contain amino acids and the proteins built up from them. That conjecture would be boosted if there were evidence that amino acids from space had actually played a part in the origins of life here. As a matter of fact, there is.

In 1982, geochemists Michael Engel and Bart Nagy at the University of Arizona made a startling claim. Contrary to earlier results, they announced having found amino acids in Murchison specimens that showed a slight excess of one handedness over the other. Critics jumped on the idea, attacking the methods the two researchers had used. Even though Engel and Stephen Macko of the University of Virginia backed up the argument with measurements of carbon isotope ratios in the amino acids that strongly indicated an extraterrestrial source, the specter of contamination left others doubtful. Fifteen years would go by before the matter was finally laid to rest. In 1997, chemists John Cronin and Sandra Pizzarello at Arizona State University showed, beyond reasonable doubt, that in Murchison amino acids of a type unknown on Earth there was an excess, between two and nine percent, of left-handed molecules over right. Some natural process in the cosmos was turning out molecules with a preferred handedness. As to what it might be, astronomers already had a pretty good idea.

The finger of suspicion pointed at polarized light. Ordinarily, light consists of waves that vibrate in all directions at right angles to the direction in which the light is traveling. But in polarized light, the direction of vibration is confined. Light that passes through polaroid sunglasses, for instance, ends up vibrating in just one direction, as if it had been made to squeeze through narrow slits in a fence. Such waves are said to be *plane polarized*. In *circular polarized* light, by contrast, the plane of vibration continuously rotates, like that of a rope being shaken by someone who's twisting his arm at the same time. The polarization can be either clockwise or anticlockwise as seen along the heading of the beam. It turns out that ultraviolet light that is circular polarized can selectively destroy left-handed or right-handed molecules, depending on the direction of the polarization. If an initially racemic mixture of amino acid molecules in space were bathed in circular polarized ultraviolet, the result would be an eventual bias of either the left-handed or the right-handed forms. The radiation would have to be intense and continuous over many thousands of years to create even a small excess. Yet given a suitable source of polarized UV, the process ought to work.

One intriguing source was proposed by Stanford University chemist William Bonner. He pointed out that neutron stars—unimaginably dense, collapsed stellar cores, left behind when some giant stars blow themselves apart as supernovae—emit circular polarized light at high energies. Maybe a neutron star had been shining around five billion years ago on the sprawling cloud of gas and dust that gave birth to the solar system? In 1998, however, astronomers came across an even more promising possibility.

James Hough at the University of Hertfordshire in England and his colleagues had built an instrument specifically to detect circular polarization. They attached it to the Anglo–Australian Telescope, near Coonabarabran, New South Wales, and then pointed the telescope at the Orion Nebula. Visible to the unaided eye as a hazy patch just below the Great Hunter's triple-starred belt, this is among the nearest places to the Sun where new stars are being formed, and a region rich in interstellar molecules. When Hough and his co-workers looked at the light from young stars after it had been reflected by gas clouds in Orion, they found that as much as 17 percent of it was circular polarized. Though their observations were made at infrared wavelengths, the astronomers were confident that ultraviolet light, which is obscured by the clouds, would be polarized in the same way. Assuming this to be the case, they did some calculations on how molecules in Orion that had left- and right-handed forms would be affected by long term exposure to such radiation. The answer: there would be an excess of one of the enantiomers of five to ten percent—similar to the range found in the Murchison amino acids.

While intriguing, the whole thing still sounds utterly fantastic. How could starlight shining on a gas cloud in space conceivably affect what kind of molecules are inside your body? And, on a purely chemical point, how could a few percent imbalance of enantiomers get blown up into the complete dominance of left-handed amino acids in living things today?

A possible solution to the second problem was suggested by the work of Kenso Soai and his team at the Science University of Tokyo. They started out with a chemical cocktail that contained a small excess of one enantiomer of the amino acid leucine. When the concoction reacted, it yielded another handed substance known as pyrimidyl alkanol. This too showed a slight enantiomer imbalance. But the alkanol then went on to serve as a catalyst in its own formation. And that's the key, because the enantiomer that had the slight edge to start with, by being able to facilitate its own production, quickly came to dominate the mixture. It's true that pyrimidyl alkanol, although organic, is not actually a biological molecule. But as James Hough pointed out, "Coupled with our recent discovery of large degrees of circularly polarized light in star-forming regions, there would now appear to be mechanisms for producing both the initial enantiomer imbalance and the amplification needed to obtain the imbalances observed in living organisms."

But there's still the question of the link between ourselves and the stars. It's one thing to have molecules floating in interstellar space, quite another to believe that they could have influenced the development of all life on this planet. How could that possibly happen?

✻

In a sense, we're all extraterrestrial. The particles in our bodies were once scattered across many light-years, and we're made literally of star dust. Every atom heavier than hydrogen of which we're composed was forged in the deep interior of a star now long dead. That is perhaps the most awe-inspiring truth that science has ever revealed, as wonderful as anything dreamed up in fiction.

From birth until senility, a star radiates heat and light that come from the conversion, inside its core, of hydrogen to helium by nuclear fusion—the building of heavier nuclei from lighter ones through energetic collisions. Only late in a star's evolution does this process make elements more elaborate than helium. When the internal temperature and pressure of an aging star climb to a critical point, the helium in the core suddenly begins to fuse to form carbon. Around the same time as this so-called helium flash, the outer layers of the star expand enormously and cool at the surface to a ruddy glow, transforming the star into a red giant. When the helium at the center of the star is exhausted, leaving behind a core choked with carbon ashes, helium fusion continues in a shell that gradually works its way outward.

What happens next depends on the total amount of matter in the star. If the stellar mass is high enough, the enlarging carbon core eventually becomes so hot and pressurized that carbon starts to fuse to form oxygen. Later, when the core is replete with oxygen, a carbon-fusing shell heads out in pursuit of the helium-fusing shell. In stars several times more massive than the Sun, oxygen may give way in turn to silicon, sulfur, magnesium— all the way up to iron. But even the most massive and highly evolved stars can't trigger fusion in an iron core because more energy is needed to bring iron nuclei together than is released by their joining.

Every element up to and including iron, then, is manufactured inside stars. Yet without some means for these new elements to be liberated, nothing productive could come of them. Fortunately, there are a number of ways by which star-processed material can be set free into the interstellar environment.

Great circulating currents in the atmosphere of a red giant dredge up freshly formed heavy elements from the star's interior and bring them to the surface. There, in the uppermost layers, the temperature is so low (less than 3,000°C) that some of the gassy products condense into solids—minuscule flecks of carbon or silicate, depending on the nature of the star.

Like celestial factory chimneys, red giants rich in carbon billow thick palls of soot: flakes of graphite, small shapeless carbon specks and, recent evidence suggests, more exotic forms of carbon molecules. The other main breed of red giant, in which oxygen predominates over carbon, yields very different solids. Oxygen in the outer reaches of these stars combines with silicon and metals to form silicates, such as magnesium silicate, which in turn stick together to make silicate grains. Both types of condensate—carbon and silicate—are shed continuously from the surface of the dying stars. If a star is massive enough, it may explode at the end of its life as a supernova, hurling the bulk of its matter, laced with every natural element (even ones heavier than iron, formed in the intense heat of the blast), into its surroundings at one-tenth the speed of light.

So there's a continuous recycling of star-processed material back into the interstellar medium: atoms and small molecules of various kinds, along with much larger (though still microscopic) silicate grains. Eventually, after many millions of years, some of this diffuse flotsam finds its way into the dense (relatively speaking) interstellar clouds from which new stars and their worlds are formed. The temperature within the clouds, 10 to 50 degrees above absolute zero, is as low as anywhere in the universe—so low that ordinary chemical reactions can't take place because the thermal motion of the particles is so feeble. Instead, all the action involves *photochemistry* (light-activated chemistry), driven by the ultraviolet radiation of nearby young stars.

Various scientists around the world have conducted experiments to show what happens to particles once they enter an interstellar cloud. Typically, these have involved comparing the way infrared light is absorbed in clouds with the way it's absorbed by different kinds of grains and ices manufactured in the lab. (Unlike ultraviolet or visible light, infrared can penetrate parts of interstellar space thick with gas and dust and so offers the best way of learning about these regions.) The wavelengths at which infrared is absorbed are like fingerprints, enabling the absorbing particles to be identified. Ice-coated silicate grains, it turns out, give the closest match between lab and astronomical measurements. So these are widely believed to be the types of grains on which most interesting chemistry within clouds takes place.

As a silicate grain enters the cold, dark interior of a cloud, it acquires an icy coating of simple molecules, such as water, carbon monoxide, carbon dioxide, methanol and ammonia. Impinging ultraviolet light then begins to break down some of the chemical bonds of the frozen compounds. Finally, because the broken molecules are held closely together in the grain's icy mantle, they're able to recombine in new ways and gradually build more complex substances. Exactly what these substances are, and what bearing they might have on the origin of life, is one of the hottest subjects in astrobiology today.

At the NASA Ames Astrochemistry Laboratory at Moffett Field, California, scientists have built equipment to mimic the chemistry that takes place in the bone-cracking cold of an interstellar cloud. Inside a shoebox-sized metal chamber, a special refrigerator and pump recreate the subzero vacuum of space. From a copper tube, a mist of simple gas molecules plays onto a bitter-cold, lollipop-sized disk of aluminum or cesium iodide that substitutes for a silicate grain. The gases instantly freeze on contact with the disk. Light from an ultraviolet lamp then bathes the newly-formed ice in a potent beam of star-like radiation, snapping bonds and stimulating reactions. Later, infrared light, shone through the ice, is absorbed at specific wavelengths by whatever chemicals are frozen inside. From the resulting infrared absorption spectrum, scientists can identify exactly what substances have been formed.

Ames researchers Max Bernstein, Scott Sandford and Louis Allamandola have been at the forefront of this work and have produced some telling results. When exposed to ultraviolet, even a very simple starting ice of frozen water, methanol and ammonia, in the same proportion thought to occur in space ice, yields a slew of interesting organics. Ethers, alcohols, ketones and nitriles all form in this space-borne equivalent of Stanley Miller's prebiotic brewery. There's also a six-carbon molecule by the name of hexamethylene tetramine, or HMT, which does something especially interesting if you add it to warm, acidified water—it forms amino acids.

The biggest molecules created by irradiating a simple ice mix contain as many as fifteen carbon atoms, and some of these larger molecules prove to have a remarkable property. David Deamer, a chemist at the University of California at Santa Cruz, has found that, when added to water, certain of the multi-carbon ice-grain substances organize themselves spontaneously into tiny rounded capsules that look strikingly like cells. When you examine these structures closely, you see they are bounded by a leaky membrane, two molecules thick. Just as in living cells, the membrane is made of molecules having hydrophilic (water-loving) heads that line up on the membrane's outer and inner surfaces, and hydrophobic (water-fearing) tails that point into the membrane's interior. Deamer saw exactly the same kind of capsule-forming behavior ten years earlier, when he added water to organic extracts from the Murchison meteorite. Both the lab- and Murchison-derived capsules fluoresce, indicating that carbon-rich, energy-capturing compounds are trapped inside—a fact confirmed in the case of the star-cloud chamber synthetics by Ames researcher Jason Dworkin.

Bernstein, Sandford and Allamandola have also experimented with more organically advanced ice mixtures. In this they were prompted by the success of other recent work that has helped to resolve a generations-old riddle. Early in the last century, astronomers noticed that the ultraviolet, visible and near (short-wavelength) infrared parts of the spectra of bright young stars are crossed by numerous dark bands. By the 1930s it was clear that the stars themselves weren't responsible for these features. The so-called diffuse interstellar bands (DIBs), of which around 200 have now been catalogued, were evidently caused by some unknown material that was spread widely throughout the general (diffuse) interstellar medium. Identifying this enigmatic stuff became the classic problem in astrophysical spectroscopy.

Over the years, almost as many theories were put forward as there are DIBs. But by the late 1990s, attention had become focused on one highly unusual suspect. Measurements of the precise wavelengths and structure of the DIBs, using sensitive equipment by astronomers such as the Dutchman Peter Jenniskens and his colleagues at the European Southern Observatory, pointed the finger of suspicion firmly at large, carbon-rich molecules as the source of the mystery bands. One of these molecules was a member of the fullerene family, otherwise known as "buckyballs." Sixty carbon atoms made up its hollow, cage-like structure, reminiscent of the geodesic domes designed by the architect Buckminster Fuller, after whom the chemicals are named. Other idiosyncrasies of the DIBs fitted well with the known spectra of polycyclic aromatic hydrocarbons, or PAHs. These are flat molecules, shaped like snippets of chicken wire, with carbon atoms linked together as tessellated hexagons and joined on the outside to hydrogen atoms. On Earth, PAHs are found in everything from automobile exhaust to charcoal-broiled hamburgers, and they are notoriously carcinogenic—an irony, considering that they may also have played a significant role in life's origins.

Both fullerenes and PAHs are highly stable molecules, robust enough to survive long exposure to ultraviolet light in the diffuse interstellar medium without being blown apart. Upon arrival within a dense interstellar cloud, they presumably condense and become frozen, along with simpler molecules, into the icy coating of silicate grains. To investigate what might happen next, Bernstein, Sandford and Allamandola ran experiments in their cloud chamber in which the starting ice mixture included napthalene, one of the few commonly known PAHs. After the usual dose of ultraviolet light, the resulting organic medley was found to contain chemicals more elaborate than those seen in earlier simulations, among them complex ethers and alcohols such as napthol. Most intriguingly, there was napthaquinone, a molecule of immense biological importance on Earth.

Quinones are found at the heart of the energy-transfer machinery within living systems. They play an essential role, for example, in converting light into chemical energy in photosynthesis. More generally, they help move energy from one part of an organism to another. And now, as the Ames experiments suggest, they may be made in the very clouds of gas and dust from which new suns and their orbiting worlds take shape.

❋

Astronomers envisage the solar system as having started out as nothing more than a locally dense clump in an interstellar cloud, tens of light-years across, which probably gave birth to hundreds of other stars. Gradually this clump pulled itself into a slowly spinning ball of gas and dust—a *globule*—no more than a light-year across. As the presolar globule shrank further under its own gravity, its rotation rate increased (just as an ice-skater spins faster as she draws in her arms), causing it to flatten out more and more. In the middle of the disk, where the density was greatest, the protosun began its final condensation to become a true star, while around it settled a dusty, pancake-shaped disk—the residual material from which the planets, moons and smaller bodies of the solar system would coalesce.

Earth took shape over a period of perhaps a few million years. Grains of dust in the protoplanetary disk collided and stuck together by chemical adhesion to make larger grains, which carried on growing in the same way. When the biggest objects had reached a size somewhere between that of a boulder and a mountain, self-gravity took over from chemical stickiness as the main process promoting further accretion. Growth then continued at a runaway pace, until the Earth and other planets had acquired more or less their present-day masses.

According to the standard view, a hundred million years or so had to pass before our planet's surface cooled sufficiently to harden as a thin crust. Never far beneath this, a sea of magma still seethed, bursting through repeatedly in numerous volcanoes and vents, discharging the gases that gradually built up to make the Earth's first substantial atmosphere. When the primordial atmosphere was thick enough with water vapor and other gases, clouds formed and it began to rain. Meanwhile, the bombardment from space went on unabated, with comets, asteroids, and meteorites of all sizes smashing into the surface at frequent intervals.

In the inner part of the solar system, close to the young Sun where the planets Mercury, Venus, Earth and Mars formed, the temperature was relatively high. Any dust grains falling into this region from the presolar nebula quickly lost their icy coatings, so that the only solid matter available for world-making was rocky or metallic. But further out, where it was much cooler, the icy coatings of grains survived intact. This frozen material became incorporated into larger objects essentially in a pristine state, still bearing the cargo of organics it had acquired in interstellar space. Icy particles that managed to avoid being swept into the gas giant planets, Jupiter, Saturn, Uranus and Neptune, stuck together to make numerous small bodies—the ice-and-rock nuclei of comets, aptly described as "dirty snowballs." Observations of comets today show their chemical make-up to be very similar to that of dense interstellar clouds.

Around the time the Earth was born, astronomers believe, many billions of embryonic comets orbited the Sun at distances similar to those of the gas giants. But that situation was short-lived. The powerful gravitational fields of the giant planets acted like sling-shots and hurled vast numbers of these icy dwarfs into new trajectories. Calculations suggest that many of the comets that formed in the vicinity of Uranus and Neptune were exiled to the Oort Cloud—an immense, spherical cometary swarm whose outer limit reaches more than a light-year from the Sun. Of those that formed closer in, near Jupiter and Saturn, a few probably also ended up in the Oort Cloud. Others were shot out of the solar system altogether, while still others were catapulted inward, into the warmer region occupied by the little, rocky planets. This sunward barrage ensured that cometary collisions with the Earth and its neighboring worlds were inevitable and numerous in the early days of the solar system. And it ensured, too, that ice and organic molecules, initially denied to these inner worlds because of the high temperatures at which they formed, were now made available. But in what quantity and to what end?

That water and various carbon-rich chemicals were delivered to Earth by impacting comets isn't widely disputed. What scientists disagree upon is the extent and influence of these extraterrestrial deliveries in the few hundred million years after the first ocean had begun to accumulate on the Earth's surface.

Some researchers have argued that comets may have brought nearly all of the water found on our planet today. This idea took a knock, however, in March 1999 when astrochemist Geoffrey Blake at the California Institute of Technology and his colleagues published results based on their observations of Comet Hale-Bopp—one of the brightest naked-eye comets in recent times. Using Caltech's newly completed Owens Valley Radio Observatory Millimeter Array, Blake's team were able to obtain a high-resolution spectrum of the very-short-wavelength radio waves coming from gas jets emitted by the comet's nucleus. Among the details this spectrum revealed was the proportion of heavy water in the comet's ice. (Heavy water contains an isotope of hydrogen, called deuterium, that's more massive than the normal variety.) The Caltech study showed that Hale-Bopp's ice is much richer in deuterium than are terrestrial oceans, indicating that the bulk of Earth's water probably didn't come from comets. On the other hand, cometary collisions almost certainly made a very significant contribution. The discovery by Lunar Prospector of billions of tons of ice on so unlikely a place as the Moon suggests that comets have delivered large amounts of water to all the major objects in the solar system. So a dual-source theory would seem to fit the bill. As University of Hawaii astronomer Tobias Owen put it, "The best model for the source of the oceans at the moment is a combination of water derived from comets and water that was caught up in the rocky body of the Earth as it formed."

At first any water that fell, either as ordinary rain or aboard colliding comets, would have sizzled on the Earth's scorching surface and been quickly turned to steam, like water droplets on a hot frying pan. But as the crust cooled, pools, lakes and small seas would have accumulated, until finally, some time after 4.4 billion years ago, the planet was covered by a single ocean, broken only here and there by scattered volcanic islands. The scene was set for the next major phase of development: the lead-up to life.

According to one school of thought, it doesn't make any difference how many objects struck the Earth during the Hadean era or what tonnage of organic stuff they had on board. Those cargoes of carbon chemicals would have been destroyed in the inferno of their arrival, rendering them useless for future biological development. Life, according to this view, was of necessity entirely home-grown.

But most scientists now disagree. Evidence has recently stacked up that plenty of organic matter both could and did survive major impacts. For thirty years, Jeff Bada ranked among those who rejected outright the notion of an extraterrestrial agency in life's origins. But then he, Scripps colleague Luann Becker (now at the University of Hawaii), and Robert Poreda of the University of Rochester found something that Bada said "blew our minds. We never expected it to be possible."

Just under two billion years ago, an asteroid about the size of Mount Everest hollowed out a 200-kilometer-wide crater near present-day Sudbury, Ontario. The crater's much-eroded remains, known as an astrobleme, are clearly visible in photos taken from space. Prospecting around this structure in 1994, Bada and his colleagues found huge quantities of buckyballs—an estimated one million tons in all—which they initially supposed had been formed from vaporized carbon at the time of the impact. Closer study, though, revealed that trapped inside the cage-like molecules were atoms of helium, an element rare on Earth but common in space. What's more, the trapped helium showed a ratio of the isotopes helium-3 to helium-4 that was distinctively extraterrestrial. The buckyballs hadn't been created on the ground at all. They'd arrived on board the asteroid—and survived the trauma of its collision. If they could do it, why couldn't other kinds of inbound organic matter?

Since that breakthrough, Becker and other researchers have found fullerenes inside a number of carbonaceous chondrites. The Allende meteorite, which fell to the ground in pieces near the Mexican village of Pueblito de Allende in 1969, has yielded buckyballs containing as many as 400 carbon atoms. The same kind of chemicals have turned up in the Murchison meteorite. Most dramatically, in March 2000, Becker, Poreda and Ted Bunch of NASA Ames reported finding extraterrestrial gases trapped inside buckyballs associated with a one-inch layer of clay that crops up in various places all around the world. This layer, the famous K.T. (Cretaceous–Tertiary) boundary, contains fallout from the asteroid that wiped out the last of the dinosaurs 65 million years ago.

On the theoretical front, too, evidence is steadily mounting that relatively fragile molecules could have been delivered safely to Earth during large-scale collisions. In 1999, Elisabetta Pierazzo of the University of Arizona and Chris Chyba published the results of computer simulations that estimated the impact survival chances of amino acids. Head-on smashes by either comets or asteroids, they suggested, would probably have incinerated these chemicals. But *grazing* impacts were a different story. According to their computer predictions, comets coming in at a shallow angle could have successfully brought significant amounts of some amino acids to the planet's surface.

And then what? Assuming this happened—that the Earth actually was inoculated during the Hadean era with all the various molecules thought to be synthesized in interstellar space—what was the upshot? These molecules would have been available to supplement whatever prebiotic chemicals were already around on the young planet. No matter where life began, whether it was on land or underground, at the top of the ocean or in the abysmal depths, the interloping particles would have eventually found their way to those places and become drawn into the genesis process. Perhaps some critical substances are hard to make under planetary conditions. Perhaps, for some reason, their production is easier on interstellar grains bathed in high-energy starlight. If so, they would have helped speed up the protobiological preparations and brought forward the day when the first terrestrial organisms appeared. As Bernstein, Sandford and Allamandola have said:

> One can imagine that a molecule, literally dropped from the sky, could have jump-started or accelerated a simple chemical reaction key to early life. If life's precursor molecules really linked up in a primordial soup, amino acids from space may have provided the crucial quantities to make those steps

possible. Likewise, life-building events taking place on the seafloor might have incorporated components of extraterrestrial compounds that were raining into the oceans. Being able to carry out this chemistry more efficiently could have conferred an evolutionary advantage. In time, that simple reaction would become deeply embedded in what is now a biochemical reaction regulated by a protein.

Call this the "facilitated-development" theory. There are a number of more radical ones. The least startling goes back to the discovery by Deamer of "membrane"-forming material in the Murchison meteorite and something very similar in lab-simulated interstellar grains. It could be that a chemical of this kind was delivered to Earth by comets, and thereafter played some role in the evolution of the first protobiological membranes. In fact, scientists have known for many years several ways of chemically making microscopic capsules that could have served as precursor cells, so this doesn't really fill in a missing chapter in the story of life's origins. What *is* new is the suggestion by mainstream astrobiologists that organic matter may have progressed significantly further along the road to life aboard the comets themselves. Bernstein, Sandford and Allamandola again:

> An intriguing possibility is the production, within the comet itself, of species [of organics] poised to take part in the life process. This "jump-starting" of the life process by the introduction of these, perhaps marginally biologically active species . . . may not be as far fetched as it would seem at first. . . . [T]here are repeated episodes of warming for periodic comets such as Halley when they approach the Sun . . . [allowing] ample time for a very rich mixture of complex organics to develop. . . . It is even conceivable that liquid water might be present for short periods within the larger comets. [L]ow temperatures . . . and gradual periods of warming and cooling might actually serve to protect the larger species in much the same way membranes are invoked for aqueous systems. Thus . . . it is quite plausible that comets played a more important active role in the origin of life . . .

"You don't say?" one can imagine Fred Hoyle and Chandra Wickramasinghe replying to this. Since the 1970s, Hoyle and his Sri Lankan-born colleague at the University of Wales have been trying to persuade the rest of the world that not only do some of the building blocks of life exist in space, *but so too does life itself.* Noting that both the size and the thermal-infrared spectrum of interstellar grains closely resemble those of dried bacteria, they were led to speculate that such grains might actually *be* bacteria. According to their theory, some of these interstellar microbes would find their way in time into cometary ice, be released by comets as they approached the Sun, and so be swept up by the Earth as it traveled round its orbit. Most extraordinary of all, Hoyle and Wickramasinghe claimed that comet-released microbes were responsible for many global outbreaks of disease, including pandemics of influenza. Not surprisingly, this idea was mercilessly attacked in orthodox scientific circles. But the notion of microorganisms drifting among the stars was far from new. Svante Arrhenius, grandfather of Scripps oceanographer Gustaf, had championed the theory of interstellar panspermia in the early twentieth century.

Arrhenius believed that life was spread around the cosmos by the pressure of starlight acting on hardy bacteria or their spores. Such spores, having drifted high into a planet's atmosphere, would be loosed by the force of the host star's radiation and propelled into interstellar space, to drift for millions of years until they eventually entered the systems of other stars, and so bring life to previously barren worlds (or perhaps compete with indigenous populations already present). The main problem with the theory—and Arrhenius was fully aware of it—was the heavy dose of potentially lethal radiation any organism would receive on such a journey. Still, he was optimistic: "All the botanists that I have been able to consult are of the opinion that we can by no means assert with certainty that spores would be killed by the light rays in wandering through infinite space." Others were less convinced, and panspermia never managed to compete with theories that put the origin of life firmly on planetary surfaces.

Today, panspermia is at least tolerated in polite company. No longer does the mere mention of the word jeopardize a promising career or invite ridicule by one's scientific peers. Versions of panspermia are discussed at astrobiology conferences, and form the subject of papers in respected journals. It's true that extreme theories, like that of Hoyle and Wickramasinghe, are still largely regarded with skepticism, but the idea of microbes being able to hop from world to world has very much entered the scientific mainstream.

Part of the reason for this growing acceptance is the discovery that some microorganisms can endure the most extreme conditions, including the vacuum of space. When NASA scientists examined the camera of *Surveyor 3*, brought back to Earth by the *Apollo 12* astronauts who landed a short distance from the probe in November 1969, they were astonished to find, living in the instrument's foam insulation, specimens of *Streptococcus mitus*, a harmless bacterium more normally at home in the human nose, mouth and throat. Since the camera was returned under strictly sterile conditions, it was clear that the microbes must have stowed away on the spacecraft before it left Earth and then survived 31 months on the Moon's airless surface. According to *Apollo 12* commander Pete Conrad, "I always thought the most significant thing that we ever found on the whole damn Moon was that little bacteria who came back and lived and nobody ever said [an expletive] about it."

Another kind of bacterium, *Deinococcus radiodurans*, or "strange berry that withstands radiation," ranks as the world's toughest. Pinkish in color and smelling like rotten cabbage, it was originally isolated in the 1950s from tins of meat that had spoiled despite supposedly sterilizing irradiation. Since then, *D. radiodurans* has earned its nickname of "Conan the Bacterium" by showing up in elephant dung, irradiated haddock, and granite from Antarctica's Mars-like Dry Valleys, apparently enjoying every minute of it. Whereas an exposure of 500 to 1,000 rads is lethal to an average person, *D. radiodurans* continues to thrive after a dose of up to 1.5 million rads. Chilled or frozen, it can handle double that amount. What's more, it achieves these feats of endurance without forming spores—the highly resistant resting phase adopted by some types of bacteria. And it isn't that *D. radiodurans* escapes damage when it's bombarded with radiation—its genetic material gets smashed into hundreds of fragments. Yet, within a few hours, it begins stitching its shattered DNA together and eventually resurrects a genome free of breaks or mutations. How and where did *D. radiodurans* learn this self-repair trick, given that there are no such harsh radiation environments on Earth? In space perhaps? Modern panspermia advocates haven't been slow in identifying *D. radiodurans* and its ilk as possible Arrhenius- or Hoyle-type interstellar travelers. A more likely theory, though, was proposed some years ago by Robert Murray of the University of Western Ontario: that the microbe's DNA-repair system is an evolutionary response to the problem of genetic damage caused by prolonged water shortage; radiation tolerance is just a side-effect. Murray's idea has recently received experimental support from the work of microbiologist John Battista and his group at Louisiana State University in Baton Rouge.

For NASA, the discovery of microorganisms capable of surviving in space is a mixed blessing. In one sense it's a problem, because extra precautions must be taken both in ensuring that spacecraft don't contaminate any worlds of possible biological interest, and also in handling any materials brought back from those worlds to Earth. The contamination issue was the main reason behind launching the *Long Duration Exposure Facility* (LDEF). In 1990, over fifty experiments that had been orbiting the Earth aboard LDEF for nearly six years were recovered by the space shuttle for analysis. Among the surprises: samples of *Bacteria subtilis* were still viable despite having being shielded from intense solar ultraviolet by only a single layer of dead cells. Calculations suggest that a millimeter of rock affords almost total protection, whereas unshielded solar UV kills 98 percent of even the toughest bacterial spores within 10 seconds.

Galactic cosmic rays (mostly fast-moving protons) are far more penetrating and potentially devastating to life. However, high-energy ones are surprisingly rare. Gerda Horneck, a specialist in radiation biology at the German Aerospace Center, in Cologne, has calculated that one spore out of 10,000 can avoid a lethal cosmic ray for 700,000 years without protection, and for 1.1 million years if shielded by 70 centimeters of rock—time enough to travel even to other stars.

There seems little doubt that some microbes could easily survive, if not an interstellar journey, then at least a passage between worlds of the same planetary system. That possibility is made all the more interesting because it's now known beyond all doubt that *rocks* can be tossed back and forth between neighboring worlds. Colliding asteroids can splash material off the surface of a planet or moon and throw it clear into space, so that at some future time it may land elsewhere as a meteorite. More than a dozen meteorites have been found on Earth that, without question, came from Mars. The proof is in the little pockets of gas these rocks contain; the composition of the trapped gases is exactly the same as that of the Martian atmosphere measured by the *Viking* landers in 1976.

Stanford geologist Norman Sleep has calculated that 10 to 100 meteorites from Mars presently strike the Earth each year—and that one out of 10,000 Martian meteorites spends less than 10,000 years in space before arriving here. In the solar system's early days, the transfer rate may have been 1,000 times greater. Nor was it all one-way traffic. Earth rocks have almost certainly ended up on Mars, though at least 20 times less frequently, Sleep estimates, due largely to our planet's higher gravity.

What if life arose on Earth and then helped seed Mars via meteorite? More intriguingly, what if it happened the other way around? The idea put forward by astrobiologist Kevin Zahnle of NASA Ames that we may all be descended from Martian microbes is starting to catch on. Certainly, the means of transport has now been established, together with the survivability of some microbes. And "ballistic panspermia" isn't just confined to Earth and Mars. It could happen between other worlds of the solar system, and between worlds of other planetary systems.

This has a couple of important implications. On the one hand, it provides a way for life to gain a foothold earlier on some worlds than if there were no biological exchange. Wherever organisms first appear around a star, that world might become the center for the diaspora of life to its planetary neighbors. In the case of the solar system, as Zahnle points out, "Early Mars may have been safer from impact sterilization than early Earth, and probably was habitable before the Earth–Moon system formed." On the other hand, the possibility of interplanetary seeding complicates the job of the astrobiologist, who will want to know if the biota on a given world are indigenous or at least partly colonial.

☀

Many questions remain unanswered about the link between organic chemistry in space and the emergence of life on a planet's surface. But that there *is* a link is now beyond doubt. Nature is not known for its generosity; it tends to make full use of whatever ready-made resources are at its disposal. Given that complex molecules are manufactured between the stars and then dropped like manna into the starting milieu for life, it's hard to believe that biology doesn't incorporate some of those interstellar products into the heart of its machinery. That conclusion is supported, as we've seen, by the similarity between some important earthly biochemicals and substances thought to exist in interstellar grains, the evidence for large-scale survivability of inbound organic matter, and the growing suspicion that the handedness of life's molecules has an extraterrestrial origin.

A cosmic connection would not only help explain some of the finer points of our biology and the extreme speed with which life sprang up here. It would also suggest that other life in the universe may share much of the same chemical basis. Not just a common carbon heritage, but the universal use of some of the same molecular building blocks, such as amino acids, sugars and quinones, is indicated (as we'll explore further in Chapter 7).

The next step for astrobiologists—the most crucial of all—is to find examples of extraterrestrial life to test this idea of chemical kinship. But where to look? With all the universe before them, in what kind of environments should scientists start their alien quest?

4

Havens, Hells, and H₂O

Several hundred meters above the ocean floor, something like a flying saucer wheels across the underwater landscape. Carried within the giant swirling vortex are chemicals, warmth, and the larval forms of living creatures, bound on an uncertain journey that began above a volcanic vent and may end with the colonization of a new vent hundreds of kilometers away. Deep *below* the ocean floor, microbial moles bore through solid rock, assimilating chemical energy as they dissolve their way along. And on the floor itself, 700 meters down, strange worms dwell contentedly in explosive methane ice, tending kitchen gardens of bacteria.

A vision of alien life? Absolutely. But these particular aliens happen to live here on Earth. They're among the mind-boggling discoveries of the past few years, encouraging the view that life may be able to adapt to all sorts of bizarre extraterrestrial regimes, perhaps including some that are on neighboring worlds in the solar system.

The notion of life as a delicate flower that needs careful nurturing under mild conditions has faded into the past—and with it the belief that life might be rare. Today, many astrobiologists have a growing sense that life is uncannily good at exploiting whatever niches the universe tosses its way. It seems almost to relish the challenge, popping up in the most outrageous places, dining happily on toxic waste. Wherever a supply of organics, liquid water, and energy exist together, scientists now suspect living things may not be far away. The question then becomes, in what sort of places do these essentials coincide. Where are the havens for life in space?

☼

Suitable energy sources seem less of a problem than before, now that thermal energy welling up from inside a world can be added to the repertoire available for would-be organisms. Basic organic matter, too, looks in good, general supply—delivered, free of charge, on the backs of comets and asteroids to the surface of every planet and moon in a young planetary system. Liquid water, though, is more of a limiting factor. True, there's plenty of ice aboard comets, so some H_2O is going to be imported whether or not a foundling world has its own native supply. But ice alone probably isn't adequate; the processes of biology as we understand them seem to depend crucially on water in liquid form. That, in turn, puts constraints on the temperatures and pressures compatible with a living environment.

But who says life needs water? Shouldn't we be thinking along broader lines, about other possibilities? Most definitely. In fact, there's been plenty of intriguing speculation, both in science and science fiction, about the biological potential of other substances, such as ammonia, methanol and sulfuric acid. These three, in particular, remain liquid well below 0° Celsius, so it's been suggested they might substitute for water in very low temperature life. Astrobiologists certainly don't rule out alternatives to water, just as they don't exclude alternatives to carbon as the backbone of organic molecules. They're acutely aware that the universe may, and probably will, spring some huge surprises on them. But they have to focus their efforts; otherwise, they'll have little chance of finding anything. And because water is so central to the kind of life we know, it makes sense to look for life in watery places.

Water seems ordinary, but it isn't. It's one of the most remarkable substances in nature. The weirdness starts when you freeze it. Unlike virtually every other chemical, water expands when it turns from liquid to solid. Being less dense in its frozen state, it floats to the surface—a crucial property as far as aquatic life is concerned. If ice sank, it would collect at the bottom of a lake or ocean, fail to melt in summer, and build up more and more in winter until the whole body of water was completely frozen, killing anything inside it.

And it goes on from there, one exceptional property after another. Water has a greater ability to hold heat than almost all organic compounds, making it a superb buffer against quick changes in temperature. Because most organisms are mostly water (only about a third of your body weight isn't), they have a built-in thermal stabilizer.

Many substances dissolve easily in water, enabling a huge variety of reactions to take place in aqueous solution. Molecules such as enzymes and nucleic acids are pushed into their specific shapes by the way electric charges are distributed on water molecules, permitting their specific biological functions. Water has a very high surface tension—the property that makes it seem as if it has a skin (upon which water striders, for instance, can "walk"). One effect of this is to encourage biological molecules to crowd together and react more rapidly, and this may have been a crucial factor in the formation of the first cells. Finally, water carries away more heat during evaporation than any other liquid—making it the perfect coolant.

As far as other worlds in the solar system go, the hunt for life is first and foremost a search for liquid water. Conversely, if scientists believe a place is bone dry, and has been for a long time, they tend to doubt its prospects for harboring anything of biological interest. Mars is a classic case in point.

✵

The study of possible life on Mars—past or present—probably accounts for about half of all the work done in astrobiology today. If you include the cost of spacecraft like *Viking, Mars Pathfinder,* and the ill-fated *Mars Surveyor 98* probes, it certainly accounts for more than half the money spent.

Serious biological interest in the planet goes back well over a century, to the time when Giovanni Schiaparelli claimed to have seen *canali* on its surface. The word translates from the Italian as either "channels" or "canals," and Schiaparelli tended to think they were natural features like rivers or rift valleys. But Percival Lowell, the wealthy Bostonian businessman, diplomat and self-taught astronomer, inclined toward a more dramatic interpretation. The canals were part of a vast irrigation system, he argued, constructed by a great civilization to transport melt-water from the Martian poles to the dry desert regions that covered most of the planet. Whatever his failings as an observer, Lowell more than made up for them with his enthusiastic speeches and writings. It hardly mattered to the lay public that most scientists regarded Lowell's artificial waterways as a gross misinterpretation or mirage. Here, for instance, is the astronomer James Keeler, speaking at the dedication ceremony of the Yerkes Observatory at Williams Bay, Wisconsin, in 1897:

> It is to be regretted that the habitability of the planets, a subject of which astronomers profess to know little, has been a chosen theme for exploitation by the romancer, to whom the step from habitability to inhabitants is a very short one. The result of his ingenuity is that fact and fiction become inextricably tangled in the mind of the layman, who learns to regard communication with inhabitants of Mars as a project deserving serious consideration . . . and who does not know that it is condemned as a vagary by the very men whose labors have excited the imagination of the novelist.

"The novelist" was clearly H. G. Wells, who, in *The War of the Worlds* published that same year, had drawn on Lowell's speculations of a superior but beleaguered race and transformed them into a dark vision of alien menace and invasion—an intellect "vast and cool and unsympathetic." No scientifically sound picture of the Red Planet could possibly compete with this.

Yet reality gradually intruded, and it became obvious to everyone that advanced life on Mars was unlikely. The canals turned out to be an illusion, the atmosphere too thin to offer much hope to the zoologist. Shifting patterns and colors on the Martian surface did, however, suggest the seasonal comings and goings of vegetation, and right up to the dawn of the Space Age there was plenty of debate about the prospects for primitive plant life. Then came the first close-up photos of the surface sent back by *Mariner 4* as the probe hurried past the planet in 1964, and a depressing scene they revealed: a crater-strewn wilderness not unlike that of the Moon. It was the low-water mark, so to speak, in Martian biological exploration. In the years since, although expectations have repeatedly waxed and waned, the dream of finding something alive or the remains of past life has been sustained.

Hope was resurrected in 1971, when the first successful Martian orbiter, *Mariner 9*, returned spectacular images of what appeared to be dried-up river channels and teardrop-shaped islands evidently sculpted by fast running water. Schiaparelli's *canali* existed after all. And they spoke of a time, long ago, when Mars was a warmer, wetter place. Its ancient atmosphere, vented like Earth's out of volcanoes and cracks in the planet's surface, would have been reasonably thick, scientists believe, between about 4 and 3.5 billion years ago. Its pressure would have been high enough to allow liquid water to exist on the surface and its thermal blanketing effect great enough to keep temperatures mild.

But Mars is a much smaller world than Earth, with a gravitational pull only a third as strong. So the Red Planet couldn't hold on to its early dense atmosphere. By three billion years ago, scientists reckon, the pressure had dropped so low that any traces of water on the surface would have immediately vaporized (for the same reason that if you heat water on the top of a high mountain, where the air is thinner, it will boil at a much lower temperature than normal.)

Today, as the song goes, "Mars ain't the kind of place to raise your kids; in fact it's cold as . . .": minus 130°C at the poles in winter, and minus 55°C averaged over the whole planet. Go to Mars and, in your excitement, make the unfortunate mistake of stepping onto the surface *sans* spacesuit and you will rapidly expire in one of two ways, depending on the time and place of your egress. If it's in winter or at night or somewhere well away from the equator, the cold will probably get you first. If you avoid death by hypothermia, you'll live only as long as it takes you to asphyxiate in the woefully thin carbon dioxide atmosphere. Either way, it will be over in a few minutes, and you'll then miss the fascinating process of your body freeze-drying until it resembles an Egyptian mummy. And a mummy, moreover, with a nasty sunburn from the intense, unscreened ultraviolet rays of the Sun.

No, Mars today is no place for a creature as fragile as an unprotected human being. But if Mars was warmer and wetter in the remote past, then perhaps life evolved there during that early, clement phase. If so, given how resilient and resourceful some organisms have proved to be on Earth, perhaps it exists there still. With this prospect in mind, scientists eagerly awaited the findings of the *Viking* mission—the first attempt by spacecraft to carry out a search *in situ* for signs of extraterrestrial life.

In 1976, the twin *Viking* landers touched down on the Martian surface, 7,000 kilometers apart, and began a series of biological experiments on soil gathered by robotic scoop. On July 28, extraordinary results started to come in from one of the experiments on *Viking 1*. Known as the labeled release (LR) experiment, it basically consisted of a culture chamber into which a small sample of Martian soil could be placed. A drop of nutrient, containing various organic chemicals labeled with radioactive carbon 14, was then added to the sample in the hope that any microbes in the soil would feed on the broth and give off radioactive carbon dioxide, which could be detected by a Geiger counter. Well, the Geiger counter went crazy and so did the scientists back at mission control. *Viking* had found life!

Or maybe not. As the days went by, the data from *Viking* grew more and more confusing. The LR and one of the other experiments onboard seemed to be giving positive life readings. But other results hinted that perhaps a purely chemical process was at work. In particular, an instrument called the gas chromatograph-mass spectrometer (GCMS), designed to detect organic chemicals in the Martian samples, found none. How could there be organisms without organics? As *Viking* project scientist Gerald Soffen said at the time, "All the signs suggest that life exists on Mars, but we can't find any bodies!"

Ten weeks later, following more tests at both landing sites, researchers were still undecided about what their results meant. Soffen summed up the general feeling: "The tests revealed a surprisingly chemically active surface—very likely oxidizing. All experiments yielded results, but these are subject to wide interpretation. No conclusions were reached concerning the existence of life on Mars."

After eight and a half months of investigations on the Martian surface, including twenty-six separate biological tests, the jury was still out—and has remained so until very recently. Most scientists rallied behind the view that some highly reactive chemical, such as hydrogen peroxide, is widespread in the Martian soil and was responsible for the recorded activity. But not everyone agreed.

Could *Viking* have actually found life? One person who's adamant that it did is Gilbert Levin, the designer and team leader of the LR experiment. He maintains that the GCMS, which was treated as the court of appeal in the affair, wasn't up to the task of detecting organic matter in the kind of low concentrations you'd expect on Mars. To back up his case, he points to validation tests that were carried out using Antarctic samples before the *Viking* mission took off. Whereas the LR unit could detect the presence of as few as 50 bacteria per gram of soil, the GCMS would fail to register with anything less than about *one million* organisms per gram—a factor of 20,000 lower in sensitivity.

Strike two against the chemical interpretation was that no one had been able to duplicate in the lab what was supposed to be happening on Mars. No one had been able to make, under simulated Martian conditions, the kind of reactive substances that were said to have caused the *Viking* results. However, that's now changed, and with it the relevance of the GCMS sensitivity. In September 2000, a team of scientists from the Jet Propulsion Laboratory and Caltech announced they'd tracked down what is probably the culprit behind those tantalizing data sent back a quarter of a century ago. They'd exposed the likely biological pretender, and simultaneously showed that life on the surface of Mars today is a virtual impossibility.

The JPL-Caltech researchers put tiny samples of the mineral Labradorite, found on Mars, in test tubes. Then they injected the tubes with a simulated Martian atmosphere, cooled them to minus 30°C, and bombarded them with ultraviolet rays at the level found on Mars. When they analyzed the samples afterward, they found that superoxide (negatively charged oxygen molecules) had formed. Superoxide is one of the most ferocious destroyers of anything organic. If this is what fooled the life detection gear on *Viking*, it wouldn't have made any difference how sensitive the GCMS was to organic molecules—there would simply have been none to find.

The big issue has become how far this natural disinfectant goes down in the Martian soil. Does the superoxide permeate to a depth of ten centimeters, a meter, ten meters, hundreds of meters? That question can be partly tackled in the lab, now that the villain of the piece has been identified. And much hinges on the answer. There's no use planning missions to scoop up a handful of surface dust to search for microbes, if superoxide has turned the top thirty meters of Mars into an organic dead zone. Deep-boring robotic moles and drills would be the order of the day. Certainly, in the light of the JPL-Caltech experiments, it seems that *Viking* never had a chance to get hold of any Martians, living or dead.

☼

Still, these are new results. In retrospect, it seems strange that the official party line was that *Viking* drew a blank, when really the casebook stood wide open. And it seems stranger that the crucial matter of the GCMS short-comings weren't aired earlier and more openly. Perhaps the combination of aridity and ultraviolet exposure at the Martian surface made life there hard for scientists to contemplate. Partly, too, NASA may have felt compelled to put before the American people and government a definitive conclusion. Having just spent $500 million on a complex, biology-focused mission, it may have thought that "We still don't know" wasn't the answer that taxpay-ers wanted to hear. But it's interesting to speculate how the NASA PR machine would have dealt with those same ambiguous results if they'd been sent back, say, in 1996. In that year, a stunning announcement was made at a major press conference by a group of scientists from NASA's Johnson Space Center (JSC) and Stanford University, to the effect that they had now indeed found evidence of Martian life. It was contained not in a sample examined remotely by a spacecraft millions of miles away, but rather conve-niently in a little chunk of Mars that had found its way to Earth.

H. G. Wells had envisioned the Martians firing their invading ships Earthward out of a great cannon. The same effect, it's now known, can be achieved by asteroids striking at a glancing angle and splattering rock and dust clear of the planet's gravitational pull. One such fragment was blasted into space from the Martian surface about 16 million years ago, drifted around the Sun through that long, lonely period, and then found itself on a collision course with a larger world girdled by great oceans. The erstwhile piece of Mars made planetfall in the Antarctic and then, some 13,000 years later, was popped into a bag by a keen-eyed biped who labeled it "ALH 84001"—the first meteorite found in the Allan Hills region in the 1984 season.

Around this same time, scientists were beginning to feel pretty confident about the Martian provenance of a handful of meteorites that had been in their possession for some time. It was only in 1993, however, that ALH 84001 was found to be from the same elite stable: the tell-tale pockets of gas trapped inside it exactly matching the Martian atmospheric composition as measured by Viking. Shortly after that discovery, the analysis began that culminated in the extraordinary claims made at the NASA press conference on July 28, 1996. A team led by JSC geologist David McKay put forward four pieces of evidence that, it said, collectively pointed in the direction of past microbial life on the Red Planet. There were globules of carbonate and magnetite (magnetic iron oxide), often found in association with bacteria on Earth. There were a variety of organic compounds that could be the decayed remains of ancient organisms. And, most visually impressive, there were what looked like tiny segmented worms—the actual fossilized remains, the team suggested, of Martian microorganisms. All of this evidence was on a Lilliputian scale, crowded within a few hundred-thousandths of an inch inside the meteorite. The "fossils" in particular were exceptionally tiny— around ten times smaller than any confirmed terrestrial microbe.

It's tempting to draw parallels between the affair of the Martian fossils and that of the Martian canals a century earlier. Lowell was attacked for his sensational and unconventional method of announcing results; so too was the NASA team. The decision by NASA's management to go public before the work was described in a peer-reviewed journal was considered a serious breach of scientific protocol. Just as most researchers rejected the idea of artificial Martian canals, so now many put forward non-biological explanations for the globules and "fossils." All of the structures and chemicals, they argued, *could* have a purely inorganic explanation. Lowell ran into trouble by drawing conclusions based on observations at the limit of resolution and beyond, and so, it seemed to some onlookers, had the NASA team.

Politics was certainly behind the hasty, headline-grabbing announcement. NASA and its chief administrator, Daniel Goldin, wanted cash for astrobiology—and after the July 28 press conference it came quickly and plentifully. The hype served its purpose in that sense; by a conservative estimate, funding for astrobiology is now at least double what it would have been without the fossils extravaganza. But aside from the hoopla, is there any scientific merit to the claims about fossilized Martian remains?

One of the key issues is the temperature at which the carbonate deposits formed. It's generally assumed these came from water, containing carbon dioxide from the Martian atmosphere, that percolated through the bedrock of which ALH 84001 was once part. But the devil is in the details. The rock itself is very old—at around 4.5 billion years, almost as old as Mars itself. It's basically a clump of crystals that formed from molten lava and not the sort of stuff in which you'd normally expect to find fossils. But ALH 84001 has had a long and violent history. It bears shock marks telling of four or five different asteroid impacts on Mars. And given the hammering it's experienced, it's hard to tell if all the supposed traces of life formed at the same time and temperature, or whether the carbonate had a different origin. In the NASA team's opinion, the water that seeped through the rock was at a mild temperature in which ordinary microbes could thrive. But there's been wide disagreement on this issue. Other researchers have come up with estimates of the temperature at which the carbonate was laid down, ranging from close to freezing to a biologically-unfriendly several hundred degrees. What about the other evidence?

The actual organic matter found in the meteorite consists of PAHs—our old friend, the ubiquitous polycyclic aromatic hydrocarbons. Again, this raises an unresolved issue: PAHs aren't associated directly with living things. They're not made by organisms and they're not found in the bodies of organisms. But they do form when living things die and decay; so, in theory, the PAHs in ALH 84001 could be the breakdown products of long-dead microbes. They *could* be. However, it isn't clear that they ever came from Mars in the first place—some scientists insist they're contaminants that got inside the rock after it landed in Antarctica. In fact, since 2000, the JSC team has acknowledged that new studies show ALH 84001—and all other Martian meteorites—have been contaminated to their very cores with identifiable Earth microbes. That doesn't rule out the possibility of extraterrestrial organics. But, even if some of the PAHs are genuinely Martian, they don't necessarily point to life. They could have come from cosmic dust particles, for instance, that arrived on Mars from space.

As for the purported fossils, they quickly acquired the more neutral label "bacteria-shaped objects," or BSOs. And they're absolutely minuscule. Most Earth bacteria are about one to two microns (millionths of a meter) long, although the smallest range down to a tenth that size, or 0.2 microns. The BSOs are tinier still—only 0.02 to 0.1 microns long—so small that some biologists think they wouldn't be capable of holding the minimum complement of biological molecules needed to run the machinery of life as we know it. Again, McKay and his associates have recently shifted their position on

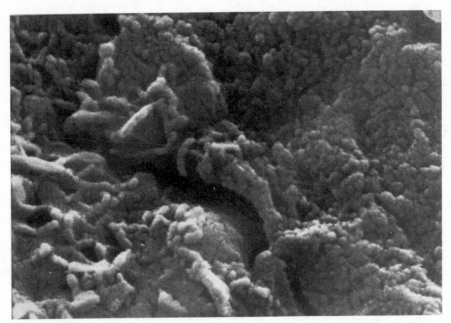

Bacteria-shaped organisms (BSOs) can be seen to the lower left in this electron micro-scope image of a section of ALH 84001. The largest are less than 1/100th the diameter of a human hair.
Credit: NASA

this score. They no longer claim that the *smallest* BSOs represent the remains of complete Martian microbes, but instead agree with other scientists that some could be inorganic crystals or ridges on mineral surfaces. It's also become clear that objects of the same size and shape as BSOs can form during the process of preparing specimens for examination under powerful microscopes in the lab.

Still, the issue of the BSOs is far from resolved. A few scientists, including Robert Folk at the University of Texas, Austin, have long maintained that there's evidence of an entire class of life-forms at a previously unsuspected tiny scale. These minuscule creatures, variously described as "nanobacteria," "nanobes" and (most recently) "nanoforms," are not only ubiquitous on Earth, according to their proponents, but can be found as fossils in a variety of Martian and non-Martian meteorites. Folk first stumbled across clusters of minute spheres, only 0.05 microns across, while studying mineral deposits from a hot-water spring in Viterbo, Italy. His claims that they were biogenic met with scorn in the research community. But, more recently, he and colleague F. Leo Lynch have found 0.1-micron balls in everything from

2-billion-year-old sediments to shower scum. Meanwhile, Hojatollah Vali of McGill University, Montreal, and his associates, have reported identifying complex biochemicals inside such structures—a discovery that could have profound implications for astrobiology. If confirmed, nanoforms might prove to be the commonest organisms in the universe, and perhaps the missing link between prebiology and cellular life.

McKay and his colleagues continue to argue that the larger BSOs in ALH 84001 could be the remains of bacteria or broken pieces of bacteria. And, in 1999, they produced new evidence of fossils inside two more Martian rocks. One of them is a chunk of a big meteorite that broke into about forty fragments over the Egyptian town of Nakhla in 1911. Local legend has it that a chunk of flying shrapnel from this explosion hit and killed a dog, which if true would be the first recorded instance of an Earthling having died at the hands of a Martian, not to mention an appalling piece of bad luck. At any rate, the new NASA images showed microscopic round and oval shapes inside tiny cracks in the Nakhla stone. What's more, these features were a few tenths of a micron long, putting them in the lower size range of some Earth bacteria. Other microbe-like forms, the NASA team announced, had turned up inside the Shergotty meteorite that landed in India over a century ago. Interestingly, the ALH 84001, Nakhla and Shergotty stones span a huge age range, from 4.5 billion to a mere 165 million years old, so if the remains inside them *are* of Martian microbes, it would suggest life has existed on Mars throughout most of the planet's history and could very well still be there today. The possibility of terrestrial contamination, however, continues to cast a shadow over these claims.

Other items of original evidence—the tiny magnetite crystals—are in some ways the most interesting. They're very small, chemically pure, and geometrically almost perfect. In 1996, when the big announcement was made, the only known sources of this kind of magnetite were some unusual terrestrial bacteria that grow the crystals so that they can orientate themselves by sensing the Earth's magnetic field. Since then, scientists have found that the crystals inside the Martian rock can be made inorganically—with one exception. About a quarter of the magnetite grains in ALH 84001 are miniature hexagonal pillars, around 0.05 micron in length. These perfect little six-sided columns, so far as is known, *can't be made in any way other than by living organisms*. Scientists largely agree that the magnetite originated on Mars. So this type of crystal, in this particular form, has to go down as the strongest evidence currently available within the Martian meteorites for past life on the fourth planet. It isn't yet proof, because there may turn out to be ways of producing such crystals inorganically (and they may be terrestrial contaminants), but it's certainly intriguing.

A lot of attention, naturally, has focused on the Martian meteorites; few planets can be sampled so conveniently. At the same time, the case for microorganisms on Mars doesn't stand or fall on the testimony of these few rocks that fortune and asteroids have cast our way. The main action is very definitely back on the planet itself, and there, the prospects for finding life seem to grow rosier by the day.

High-resolution photos and other data sent back by *Mars Global Surveyor* (MGS), which entered orbit around the Red Planet in 1997, have bolstered the case for extensive ancient surface water. They've helped to convince some scientists that a great ocean once spread across what are now the northern lowlands of Mars. The shorelines and floor of this long-gone alien Atlantic, together with the sides of water-cut canyons and channels, will be prime targets for future exopaleontologists. The layers of sediment that built up in these regions may well yield fossilized remains of primitive creatures that lived on Mars more than three billion years ago.

Most of the water that once existed on the surface is thought today to be locked away, some at the poles but most underground. One popular idea is that there's a thick layer of permafrost, deeper even than that at the Antarctic. Below this, perhaps kilometers down where the temperature is higher, the pores of subterranean rocks may be filled with liquid water. If such Martian aquifers exist, they may support an extensive deep, dark biosphere like that on Earth, harboring microbes that retreated into the planet's interior as surface conditions became too cold and dry to support life at higher levels. On the face of it, water on or near the surface today seems very unlikely. In an atmosphere a hundred times thinner than Earth's, it would boil away in an instant—wouldn't it?

June 2000. The newswires are buzzing with gossip that NASA is about to announce something big—so big that the White House has been notified. There's talk that MGS has found traces of liquid water in deep parts of the huge canyon system on Mars, Valles Marineris—the Valley of the Mariners. Whatever the story is, it's to be published in the journal *Science* on June 28. But the rumor mill is running wild and, in an unprecedented joint decision, NASA and *Science* agree to put an early end to the speculation. On the morning of June 22, in separate press conferences held an hour apart, the news is released to the world. Mars and water and MGS are indeed the stars, but the story is a little different from what most people had anticipated. The orbiting

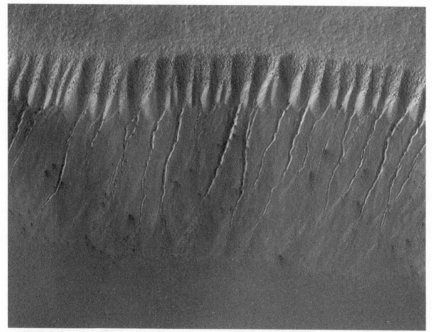

Deep, v-shaped channels, believed to be "gully washers" on the walls of a pitted region on Mars at a southern latitude of 70 degrees. This image, taken by *Mars Global Surveyor* on July 14, 1999, shows a region 2.8 km wide by 2.1 km high.
Credit: NASA/JPL/Malin Space Science Systems

probe hasn't actually spotted liquid water. In scores of photos showing detail as small as a mini-van, MGS has revealed extraordinary, stream-like features cascading down the sides of more than a hundred craters and canyon walls. All of the features consist of a deep, narrow channel with a collapsed region at the upper end, on the crater or valley rim, and a fan-like area of debris at the lower end. If geologists saw such a structure on Earth they wouldn't hesitate in identifying it: a gully-washer, caused by water erupting in a sudden flood just below the surface, high up on a steep slope, undermining and collapsing the ground there, and then racing downhill carrying material with it, before spreading out on the gentler slopes below and depositing its load. The MGS pictures were hard to interpret in any other way.

The astonishing thing is this: the Martian gully-washers—if that's what they are—were formed very recently in geological terms. None is thought to be older than about a million years, and any of them, for all we know, might have been made last week. They represent powerful evidence that liquid water still exists on Mars at shallow depths—a stunning breakthrough in astrobiology. Here, around these gullies, is a place for scientists to concentrate their efforts in the search for nearby extraterrestrial life. They no longer have to contemplate drilling perhaps through thousands of meters into the Martian bedrock to look for deep-dwelling refugees. If organisms still exist on Mars today, where are they more likely to be than near this unexpectedly convenient water source?

But how did the gullies form? What is it that, against all expectations, allows water to be in a liquid state within perhaps a hundred meters of the Martian topsoil? To compound the mystery, these features are not found in the relatively warm equatorial regions, where the daytime temperature can climb well above the freezing point of water. They seem to be most common in the southern hemisphere of the planet between latitudes 30 degrees and 70 degrees, where it's always numbingly cold. What's more, they tend to be found in places that are well shaded from the Sun, where the surface temperature rarely rises above that of an Antarctic winter night. One possibility is that the water is heavily laden with salts which could drop its freezing point to as low as minus 60°C. That needn't be a problem for life. Salt-loving microbes known as halophiles have been discovered on Earth in places like Utah's Great Salt Lake and in so-called saltines that can be up to ten times more briny than ordinary seawater. Whatever the explanation, astrobiologists have found a new, exciting focus for their explorations.

In December 2000 came more extraordinary news. MGS photos, of many different regions of the planet, showed what seem to be fine-grained minerals deposited in horizontal strata—the hallmark of sedimentary rocks. The best explanation available is that these layers formed in lakes and shallow seas, and are therefore the best places to look for Martian fossils.

The Red Planet looks more inviting to astrobiologists than ever. Yet the net of biological possibility extends beyond Mars, and beyond the asteroid belt, into another, more unfamiliar part of the Sun's domain.

✴

Huge planets with extravagantly thick atmospheres of hydrogen, helium, methane, and ammonia hardly seem ideal homes for anything remotely like life. However, that didn't stop Carl Sagan and his colleague Edwin Salpeter from speculating about "abundant biota" in the Jovian cloud-tops in 1977, just before the twin *Voyager* probes arrived to survey them. Sagan and Salpeter envisioned an ecology somewhat like that of terrestrial seas, in which "sinkers," "floaters" and "hunters" took on roles similar to those of plankton, plankton-eaters and marine predators. All of these creatures would be immense gas bags that propelled themselves along by pumping out helium, the hunter variety growing perhaps so large that they would be visible from space.

Sadly, though perhaps not so surprisingly, the *Voyagers* glimpsed nothing compellingly biological jetting around in the Jovian cloud-tops. But what they did see, as they flew by, makes its own chapter in astrobiology: a moon peppered with active volcanoes, and another covered by ice whose surface looked suspiciously young.

The volcanic moon is Io, similar in size to our own Moon but sensationally different in appearance—its gaudy oranges, yellows, reds and browns suggestive of a cosmic pizza. The colors are those of sulfur and sulfur compounds that blanket the surface, having been shot out of enormous volcanoes. Eleven were seen erupting at the time the *Voyager* probes flew by, out of a total active volcano population now put at around 300. In fact the entire moon is a vulcanologists paradise. It is always and everywhere volcanically unstable for one simple reason: tidal heating.

Because it follows an orbit made slightly elliptical by the pulls of Jupiter's other big moons, Io moves alternately closer to and further away from its monstrous parent planet as it completes each circuit. That results in sizeable changes in the gravitational tug it experiences from Jupiter which, in turn, causes a lot of flexing and relaxing of Io's interior. Just as repeatedly stretching and releasing a rubber band makes the band warm, so the continually varying forces on Io build up a huge thermal energy store inside the tortured moon. The heat escapes in the form of the worst case of tectonic acne ever seen.

Of the four large "Galilean" moons of Jupiter (so named for their discoverer), Io is the nearest to the central planet and therefore the most seriously inconvenienced by tidal stress. Next out is Europa, with an appearance completely different than that of Io, but hardly less bizarre. From a distance it looks like a cracked and badly stained cue ball. But up close it's another and far more complex story. With the arrival of the *Galileo* spacecraft at Jupiter in December 1995—the first probe to go into orbit around the planet—scientists had the opportunity to study Europa in great detail over a period of several years. And everything they have seen in that time tends to confirm what was already suspected following the *Voyager* encounters: beneath Europa's icy outer crust may lie a deep, watery ocean.

A 20 km wide region of ridged plains on Europa taken by the *Galileo* spacecraft on December 16, 1997 at a range of 1,300 km. The many parallel and cross-cutting ridges commonly appear in pairs, with dark material in between.
Credit: NASA/JPL

The lack of cratering is the first clue. An old surface is heavily pock-marked with large and small impact craters, like battle scars picked up over an aging warrior's long lifetime. Europa has very few craters, and hardly any large ones, suggesting that its present landscape is at most only a few tens of millions of years old. And what a landscape! Nothing like it exists anywhere else in the solar system: an elaborate tapestry of fractures, ridges, bands and spots. The fractures, running everywhere, are cracks in the icy coating. Close study of these cracks—their shape and length—reveals that they could only form if the surface rose and fell by many meters every day. And the only way *that* sort of impressive tidal ebb and flow could happen is if there were something very flexible underneath the hard outer shell.

Europa's ridges, similarly ubiquitous, come in pairs with a narrow valley in between. That, too, suggests something has pushed up from below—liquid water perhaps, or slushy glacial ice. Oddly, the fractures and ridges are often repeated, side by side, resulting in a patchwork-quilt of parallel lines, as if the stress pattern had swept across the surface over time. It appears that Europa's surface rotates slightly faster than its interior—an unusual state of affairs best explained if there were a subsurface ocean that acted as a hydraulic bearing.

Other signs hint at a soft, mushy or fluid layer hiding under the hard icy exterior. The few large impact craters visible on Europa are surrounded by many concentric rings, as if a rock had been thrown into a pond and the resulting ripples had suddenly frozen solid. From the structures of these craters, scientists estimate the weak layer below the ice starts at a depth of six to fifteen kilometers. That agrees roughly with calculations on how tidal heating would be expected to melt the surface icy coating from below. Putting all the known facts together, the most popular theory of Europa's interior today is that there's an iron core surrounded by a deep rocky mantle, on top of which sits the crust of H$_2$O, liquid for the most part, with an icy topping a few kilometers to a few tens of kilometers thick.

So much water invites the question: Is there also life? There's an energy source—tidal heating—because that is what has supposedly melted the ice. Only one crucial ingredient remains: organic matter. Colliding comets and asteroids must inevitably have brought in some. But the most interesting possibility is that the same tidal stresses that (in theory) maintain the ocean may also inoculate it with chemicals through hydrothermal vents. Incessant global flexing should heat the moon's rocky mantle and lead to vulcanism similar to that found on Earth's ocean floors. And so it may be that, as on our own planet, oases of life have sprung up in the warm mineral rich waters of deep-sea vents on Europa. There's even been speculation about more advanced Europans—big marine predators or whale-like filter feeders, for example, plying the gloomy subterranean sea. But life here, if it exists at all, could be restricted by the fact that its environment is sealed off. On Earth, oxygen from photosynthesis dissolves in the seawater and finds its way down to the ocean depths. But on Europa, chemical energy may be in much shorter supply, imposing limits on the complexity and diversity of life.

Or perhaps not. Richard Greenberg and his team at the University of Arizona believe that the ridges on Europa are formed when cracks penetrating down through the ice crust are occasionally pulled open by tidal stress, allowing water from below to gush to the surface. Quickly this water freezes to form a plug that fills the top few meters of the crack. Later, when tides squeeze the fissure closed, the ice plug is pushed up to form a raised ridge. If the crack pulls open again, the new ridge can be split in half and another ridge form within it. The result, Greenberg explains, is that liquid water is pumped up and down in the crack every few days, for perhaps tens of thousands of years—until the creeping rotation of Europa's surface causes the focus of tidal stress to move on and the crack permanently freezes over. This theory has important implications for future biological surveys of the moon. It opens up the possibility of finding organic material and perhaps even frozen life-forms on the surface.

Perhaps more significantly, as Chris Chyba has pointed out, water that is pumped back down could carry into the underground ocean various chemicals that are formed in the topmost layer of ice by particle bombardment from Jupiter's intense radiation belts. These chemicals would include both radiation-formed oxidant chemicals such as sulfur, hydrogen peroxide and free oxygen itself, and organics dumped on Europa by impacting comets—a potentially crucial source of food and energy for whatever might dwell in the watery ocean below.

Europa may not be the only habitable moon in the solar system. Callisto, the outermost of the Galilean satellites, also shows signs of having an underground ocean. Ganymede, the biggest Jovian moon, and even Triton, circling faraway Neptune, are on the same short-list. In fact, biologically promising bodies of hidden water may be common features of the big moons of gas giants, in our planetary system and beyond.

☼

One of the most hauntingly beautiful images of space is not a photograph. It is a painting—perhaps the best known work of that inspirer of many a present-day astronomer and space scientist, the American artist Chesley Bonestell. It shows what Saturn might look like from the surface of its biggest moon, Titan. At least, it gives a 1950s version of that speculative scene. The sky is blue, because even then it was known that Titan is the only moon to have a substantial atmosphere, and in that sky hangs the sixth planet, with its gorgeous ring system dominating the scene.

The modern truth about Titan is a little different but no less intriguing. It does indeed have a thick atmosphere, but it is neither blue nor transparent. Saturn's big moon is shrouded by a dense orange smog. It's surface is completely obscured, so that whatever is down below represents the biggest expanse of unseen astrobiologically relevant real estate in solar orbit.

Ideas about Titan are presently built more on surmise than solid fact. But this we do know: the atmosphere—denser than Earth's—is mostly nitrogen with some methane and hydrogen. The orange haze is caused by sunlight breaking apart molecules and allowing them to recombine into more complex substances with fall hues. Plenty of interesting organic chemistry is surely taking place on Titan, but how far it has advanced in the direction of life, we have as yet few clues.

Infrared observations by the Hubble Space Telescope and by the 3.6-

meter Canada-France-Hawaii Telescope atop Mauna Kea in Hawaii have glimpsed *something* on the big moon's surface—bright and dark patches, a couple of hundred kilometers across. They might be seas or continents or both. The suspicion is that most of the surface is made of ethane, in both its solid and liquid state. Titan may rain methane and smell generally like an oil refinery, with all sorts of hydrocarbons wafting in the air. But biology or pre-biology? That's an interesting question, because the conditions and the chemical blend on Titan are very different from any we've previously encountered.

Of course, Titan is a horribly cold place—minus 178°C may be the best it can normally muster. In that kind of icy chill, chemical reactions either don't take place at all or happen painfully slowly. Yet, over a few billion years, who knows what might have built up? There are some energy sources to occasionally push things along: the original internal heat of the moon, a modicum of tidal warming from Saturn, sunlight, possibly some volcanic activity, almost certainly the occasional impact. But what sort of chemistry goes on in a place like this? How sophisticated can it become? From three-quarters of a billion miles away, telescopes have spotted twenty different compounds in Titan's atmosphere. A few more—a few thousand more, perhaps—will show up when we can take a closer look.

Yet Titan has no free oxygen. It has no water that we know of. Under those circumstances how far can a world progress in the direction of life? Perhaps the ice brought in by colliding comets—which may temporarily melt on impact and make oxygen available too—is an important factor. Titan is very unlikely to be inhabited. It would be surprising even to find small proteins there. But Titan will help us calibrate how far prebiological chemistry can develop in an environment that, by modern Earth standards, is alien in the extreme.

✵

Two decades ago, it seemed as if our own planet might be the only place in the solar system of interest in the search for new life. Now we know of at least half a dozen other worlds on our cosmic doorstep to which astrobiologists would love to make a field trip. But this is just the beginning. Even before the news broke of the Martian fossils, astronomers had announced a discovery that will have far greater long-term implications in our quest for fellow life across the cosmos. They had found the first planets of other Sun-like stars.

5

Strange New Worlds

Each pinpoint of light in the night sky is a sun, and each sun is potentially at the center of a planetary system like—or perhaps not so like—our own. All told, there are some 400 billion stars within our galaxy. How many have planets? How many planets are potentially habitable? And on how many has life actually taken hold?

As soon as astrobiology casts its net beyond the solar system, it runs into the daunting problem of interstellar distances. Imagine the Earth to be hollow and a dozen oranges scattered fairly evenly around the cavernous void inside. That gives some idea how far stars are apart in relationship to their size. Even the nearest star is almost unimaginably remote—10,000 times farther away than Pluto. If it were accessible by interstate highway, you would need about 50 million years to drive there. And that's the *nearest* star. Most of those specks of light in the night sky are tens or hundreds of times farther away. How can we possibly learn about any strange new worlds that might accompany them? And, assuming there are plenty of worlds out there, what determines whether they can support life, and of what kind?

☼

One way to assess the biological potential of other stars and their worlds is to think about hypothetical planetary systems. Following their favorite rule of thumb, "Where there's water, there might be life," astrobiologists like to ask: Where around a star could water exist in its liquid state? The answer leads to the idea of a *habitable zone* (HZ). Put an Earth-sized planet in a star's habitable zone and its temperature would be moderate enough for it to support surface water and therefore features like rivers, lakes and oceans. A straightforward equation in physics, the Stefan-Boltzmann equation, gives a rough and ready estimate of a star's HZ. This is the oldest and simplest way of thinking about a stellar system's capability of supporting life. Although, as we shall see, the concept has grown less relevant of late, it's still worth exploring.

To use Stefan-Boltzmann, simply feed in a value for the star's luminosity—how much energy it radiates into space every second—plug in the temperature range for liquid water, and the equation does the rest. Well, almost.

The plain Stefan-Boltzmann formula applies to a completely black surface that always directly faces the central star. But planets aren't black. The Earth, for instance, with its clouds, polar caps and large expanses of water, reflects back into space more than a third of the solar radiation that falls on it. What's more, because the Earth rotates relative to the Sun like a chicken on a spit, the heat it does absorb is spread over the whole of its spherical surface. Allowing for these factors, Stefan-Boltzmann predicts an HZ for the Sun that extends from about 0.7 to about 1.3 times the size of the Earth's orbit, putting us right in the middle of the comfort belt as we might expect for a biological Shangri-la. What about other stars? Where do their habitable zones lie?

All stars in their prime make heat and light in the same way, by fusing hydrogen into helium. But they vary enormously in the rate at which they do this, and therefore in their total brightness. Some stars are the stellar equivalent of glow-worms. Known as red (or M-type) dwarfs, they're not only the dimmest but also the coolest, least massive, longest-lived and commonest of normal stars. The HZ of a typical red dwarf lies entirely within a radius about one thirtieth that of the Earth's orbit. That's to say, in broad terms, unless a planet circled within about *five million kilometers* of its red dwarf host, it wouldn't keep warm enough to harbor liquid surface water. Right away, that puts a question mark over whether life can be associated with red dwarfs at all. Scientists don't know even if it's possible for planets to form so close to their host star. It might be that there is never enough material near the center of the dusty disks from which planets are made for worlds to accumulate there. This is a significant issue for astrobiology, because fully three-quarters of all stars are red dwarfs.

For the moment, though, let's assume that worlds *can* exist within the habitable zones of these glow-worm stars. Then other factors come into play. Lighter bodies that circle closely around more massive bodies tend to fall into what is called "captured rotation." You can see examples of this all over the solar system, especially among the inner satellites of Jupiter and Saturn, but it's most obvious in the case of our own Moon. We're compelled to look always upon the same face of the Moon from Earth because the Moon spins on its axis in exactly the same time it takes to complete one orbit. What has happened is that, over many millions of years, the Earth's pull of gravity has acted as a brake, slowing its satellite's spin so that now it is locked in step with its orbital motion. The same would happen, scientists are fairly sure, to any planet that moved within the tiny habitable zone of a red dwarf.

Captured rotation would make for unusual surface conditions. One hemisphere would bask in the heat and blood-red glare of a sun that never sets; the other would be in eternal night. From the point of view of life, that might pose problems if the climate extremes were too severe; although even if they were, there would presumably be a narrow twilight zone between baking day and frigid night where conditions struck a happy medium. On the other hand, if a red dwarf world had a thick atmosphere, the entire surface might be rendered habitable—the continuous circulation of air between the sunlit side and the dark side would spread heat more or less evenly over the whole planet.

The other factor that might pose an obstacle to life around red dwarfs is that these stars are notorious for having stellar flares. In a matter of seconds, their ultraviolet or X-ray output can shoot up many thousandfold before dying away again over the next few minutes. That could make life problematic. Or it might encourage the evolution of organisms that can cope with sudden, sporadic variations in radiation exposure—that thrive under such circumstances, and perhaps even require them. At this stage, we've no way of knowing.

As we step up the stellar luminosity scale and enter the realm of orange (or K-type) dwarfs and then of yellow (or G-type) dwarfs, like the Sun, so the habitable zone moves outward and widens. In the Sun's case, as we've seen, Stefan-Boltzmann predicts an HZ that runs from about 0.7 to 1.3 times the size of the Earth's orbit.

But something is wrong with these figures. They put Venus (0.72 times the Earth's distance) just inside the inner edge of the Sun's habitable zone, while Mars (1.52 times the Earth's distance) is beyond the outer margin. Yet Venus is as uninhabitable as you could imagine—an inferno with a surface temperature that never dips below about 460°C. If you could survive on its surface long enough to look around, you'd see that the rocks actually glow with heat. On the other hand, scientists are still open-minded about the possibility of finding microbes on Mars, even though Stefan-Boltzmann puts it out of biological bounds. Surely, our HZ calculation has gone haywire.

Not really—the basic calculation is sound. It's just that other factors, in addition to the amount of energy received from the central star, affect a planet's climate. The most important of these, in the case of Venus, is how well the atmosphere retains warmth. Some gases, like carbon dioxide, sulfur dioxide, and methane, act like the glass in a greenhouse. They let incoming sunlight pass through, but after this has been taken up and reradiated by the ground at longer, thermal wavelengths, the greenhouse gases absorb it and warm up as a result. That raises the whole global temperature. Venus has a massive atmosphere, with more than ninety times the surface pressure of Earth's, and most of it is carbon dioxide. Consequently, the planet is in the grip of a perpetual, rampant greenhouse effect. Without this phenomenal ability to retain heat, conditions would have been a lot more congenial to the development of flora and fauna. Instead of a lava-strewn wasteland, Venus might have been a steamy jungle-world with lush vegetation—especially at higher latitudes—and perhaps a diversity of animal life. So what went wrong?

At one time, the prospects must have looked good for Venus. Four and a half billion years ago, you might have guessed it was going to progress along the same lines as the Earth. It's a sister to the Earth in size, its crust was no doubt pumping out much the same kind of primordial atmosphere, and it's even possible that life started to develop on the second planet sometime during its first half billion years. But while life may have been developing on the ground, Venus's dire fate was being sealed in the atmosphere above.

The exact sequence of events isn't certain. One theory is that Venus never had any surface water and suffered from runaway greenhouse heating almost from day one. An alternative, favored by James Kasting of Pennsylvania State University and his colleagues, who've spent many years looking at the problem, is that the temperature remained low enough early on for oceans to exist. Yet after the first few hundred million years, Venus lost its water and, as a direct result, ended up with an immense atmosphere of carbon dioxide. Biologically, it was a lethal double whammy. But why did it happen on Venus and not on Earth?

Carbon on our own planet journeys through a complex cycle that keeps a large fraction of it sequestered inside the Earth for millions of years. When water evaporates from the surface, it cools, condenses and forms clouds at a height of no more than a few kilometers, then falls again as rain. While in the air, it combines with carbon dioxide to form a weak acid—carbonic acid. Once on the ground, the rainwater physically wears away rocks that contain calcium-silicate minerals. At the same time, the carbonic acid chemically attacks the rocks, releasing calcium and bicarbonate ions (charged particles) that are carried by streams and rivers into the ocean. Plankton and other ocean creatures then use these ions to build chalky shells of calcium carbonate, and when they die they settle to the seafloor, where their shells form carbonate sediments. (Such sediments would build up anyway, even in the absence of life, through inorganic reactions.) As millions of years go by, the seafloor moves like a conveyor belt, carrying the carbonate sediments to the margins of the continents. There it dives under the continental land masses where it is heated and pressurized, causing the calcium carbonate to react with silica (quartz). In that process, silicate is reformed and carbon dioxide gas is released. This gas reenters the atmosphere through cracks in the Earth's crust—at the mid-ocean ridges or, more violently, during volcanic eruptions.

On Venus, this carbonate–silicate cycle got jammed at an early stage. The root of the problem was that the very warm surface encouraged a high rate of evaporation. So much water vapor wound up in the atmosphere that, instead of condensing as rain at low altitudes as happens on Earth, it rose to a height of 100 kilometers or more. And that was disastrous. At those high altitudes, the water molecules were split apart—photodissociated—by solar ultraviolet rays into hydrogen and oxygen. The hydrogen atoms, being very light and therefore fast-moving, simply escaped into space. And as the hydrogen leaked away, so, effectively, did Venus's water. With an end to rainfall, there was no way to wash carbon dioxide out of the atmosphere. The entire carbon cycle shut down, including the carbon-silicate cycle that sequestered these minerals within the crust. The greenhouse thermostat was turned up to maximum and left there. If Venus hadn't been sterile before, it was doomed to lifelessness now.

And yet, for a while, it had *oxygen* in its atmosphere—the oxygen left behind when the hydrogen from the split water vapor headed into space. That raises an important point, because we tend to think of free oxygen as coming exclusively from plants and other photosynthetic organisms. As far as we know, this is the only way a planet could maintain a substantial oxygen-rich atmosphere over very long periods. Being highly reactive, oxygen quickly combines with other substances, so that unless it's replenished it disappears quite soon as a free gas. That's exactly what happened on Venus: the oxygen combined with materials on the surface, and after a few hundred million years hardly a trace remained in the atmosphere. Still, the case of Venus shows that oxygen *can* come about by purely non-biological means. So if we're trying to tell whether a planet supports life by analyzing the composition of its atmosphere, we can't draw firm conclusions from the presence of oxygen alone. Finding oxygen would be exciting, but we would have to look for other chemical signs of life as well. Otherwise, we might simply be seeing a Venus-like world during the early phase, when it's losing water by photodissociation.

Venus's inability to recycle its carbon dioxide is a direct outcome of having lost its water. But there's another aspect of the story that has a critical bearing on the habitability of worlds in general. If you ask a planetary geologist to point to one process above all others that determines if a world can support life-as-we-know-it on a long-term basis, the chances are that he will tell you this: *plate tectonics*. At first, that seems surprising. What does geophysics have to do with biology? Yet, it turns out that the sliding of crustal plates and their subduction—when oceanic plates delve down under lighter continental plates—is the single most important regulator of global climate. Plate tectonics is at the heart of the carbon dioxide recycling loop. Switch off the conveyor belt of the crustal plates and you switch off the most important means of regulating the amount of carbon dioxide in the atmosphere. Then one of two things happens. Most of the planet's carbon dioxide ends up either in the atmosphere (in which case a massive greenhouse effect, as on Venus, makes it too hot for life) or locked in the ground (leading to a Mars-like freeze-up).

At first glance, you might think the only thing necessary to keep plate tectonics rolling along is internal heat. As long as enough heat wells up from the core, the rocks immediately below the crust, in the so-called asthenosphere, are kept soft and semi-fluid and so can support convection currents. These rising and falling circulations are what drive the motion of the crustal plates. But a few years ago, Caltech planetologist David Stevenson realized that something was missing from this picture—and that Venus held the key.

Being similar in size to Earth, Venus has roughly the same amount of internal heat and it almost certainly has volcanoes that have been active in recent times (and may well be active today). But volcanoes are a more general phenomenon than plate tectonics. On Earth, you find them near where oceanic and continental plates meet—for example, in the Cascades of the American Pacific Northwest. And you also find them in other places, like Hawaii, where they push up through weak spots in the crust. On Venus, all the volcanoes are of the latter kind, known as shield volcanoes, because although Venus is internally hot, it doesn't have plate tectonics. Why? Because it doesn't have liquid water. Stevenson realized that water is the essential lubricant that allows crustal plates to glide freely on top of the hot mantle rocks below. As oceanic plates on Earth sink into the asthenosphere, they carry water into the mantle, which lowers the melting point of the mantle rocks and thereby softens them to the point at which they can serve as low-friction bearings for the plates above. Without liquid water, the roller mechanism would seize up and plate tectonics grind to a halt, as it has done on Venus.

Understanding why some planets turn out Earth-like while others don't isn't just a question of pinning down one or two isolated factors. The climate, the makeup of the atmosphere, the amount of heat coming from the central star, the size of the planet, what's happening on and below the ground—all these are linked together. The terrestrial life-support machine runs on interlocked cycles: the hydrological cycle, the carbon cycle and the recycling of carbon dioxide which is part of it, the nitrogen cycle, the steady slip-sliding of the oceanic crust. Even clouds exert a regulatory effect. Consider this: if it weren't for greenhouse heating, Venus might actually be *colder* than the Earth. Its dense, white, globe-encircling clouds of sulfuric acid reflect 80 percent of the incident solar radiation back into space. On Earthlike planets, clouds are part of the climatological/geological feedback mechanism. If the surface gets too hot, more water evaporates, which increases the cloud cover, so more of the incoming energy from the Sun is reflected back into space, which helps cool the surface. At the same time, extra water vapor in the atmosphere helps wash out more carbon dioxide, and so reduces greenhouse warming.

These additional factors—the interlocking cycles of atmosphere, surface water and plate tectonics—affect the size of the Sun's habitable zone. Put the Earth where Venus is, and in time it would become lifeless. At 0.7 times the Earth-Sun distance, Stefan-Boltzmann says the surface temperature of a planet like our own, determined purely by solar radiation influx should be marginal for liquid water. But as we've seen, photodissociation at that distance drains the water away, and ultimately creates a massive carbon dioxide atmosphere. If you want to think about it another way, the lack of water puts a stop to plate tectonics. The result is the same: massive greenhouse heating and an environment in which any kind of life that we can realistically imagine is taken off the drawing board. Water loss by photodissociation, Kasting found, moves the inner edge of the HZ out to about 0.95 times the Earth–Sun distance, putting us just a few percent within the safety limit.

What about the outer margin of the habitable zone? Here again, the decisive factor is the ability of a planet to keep recycling its carbon dioxide and water. Whereas a sun-drenched Venus-like world ends up with most of its carbon dioxide in a huge sweltering blanket of an atmosphere, a planet that's too far from its star's warming rays has the opposite problem: most of its carbon dioxide and water is eventually frozen on or under the surface. And, again, it hinges on the failure of the key negative feedback loops that would otherwise keep the environment suitable for life. As the carbon dioxide goes out of the atmosphere, for example, the greenhouse effect diminishes, which leads to further global cooling. Kasting has suggested that this will happen to an Earth-sized planet moving around a Sun-like star if its average distance is more than 1.37 times the Earth–Sun distance.

But other factors complicate the picture further. Mars, for example, is *not* Earth-sized. The main reason Mars cooled down so quickly isn't that it receives too little sunlight but that, being small, it lost most of its internal heat early on. Any primordial Martian plate tectonics quickly ground to a halt, and thereafter more and more of the Red Planet's stock of carbon dioxide was permanently locked in the crust.

At the same time, it's clear that, long ago, Mars had plenty of liquid water on its surface. How could that be if it lies outside the habitable zone? Mars averages about one and half times Earth's distance from the Sun, whereas the Stefan-Boltzmann equation predicts an HZ outer limit of 1.3 and Kasting (allowing for greenhouse gas heating) pushes it out to 1.37 times the Earth-Sun distance. By that reckoning, the fourth planet should always have been a cold, desert world.

It's true that even a thick Martian atmosphere of carbon dioxide couldn't supply all the greenhouse heating needed to keep the surface warm enough for liquid water to exist. However, scientists in the United States and France have recently theorized that young Mars had another way of cranking up its thermostat—dense, frozen clouds of carbon dioxide. "In its early days," claims Raymond Pierrehumbert of the University of Chicago, "Mars was the white planet rather than the red planet." These carbon dioxide clouds would actually have reflected 95 percent of the Sun's rays back into space. But the bit of radiation that did get through, after being emitted as infrared warmth from the ground, would have been bounced right back down again by particles in the ice cloud layer. Calculations suggest that the infrared mirroring effect of the carbon dioxide crystals could have helped maintain the surface temperature at a balmy 25°C. Moreover, the ice clouds would have served as a very effective ultraviolet shield, making Mars a more benign place in those days than the surface of the Earth. Quite conceivably, whereas terrestrial life had no choice but to remain underground or in the deep sea where there was protection from the UV flooding through the (then ozone-less) atmosphere, early Martian life—assuming there was any—didn't need to hide. On the other hand, the same clouds that blocked the UV would also have allowed very little visible light to pass through, so that the surface would have been permanently gloomy. Low-light levels wouldn't have made the evolution of photosynthesis worthwhile—and that has some interesting consequences. If Mars had been the size of Earth, it might have retained a dense atmosphere, and so stayed warmer and wetter for a much longer period. But, even if it's surface had remained habitable to this day, it might never have produced advanced life. With a frozen cloud deck so effective at turning light away, there would probably have been no photosynthesis, and therefore no production of a thick oxygen atmosphere. And without oxygen it's hard to imagine the appearance of anything in the way of complex life. This, however, is just one possible scenario. As we'll see, Christopher McKay of NASA Ames has argued that, because of limited tectonic activity, Mars might have had an early oxygen-rich atmosphere in which complex life could have evolved many times faster than it did on Earth.

We begin to get some idea of the bewildering number of variables and the complexity of the interlocking mechanisms involved in the question of planetary habitability. For the most part, astrobiologists have concentrated on the evolution of Earth-like worlds. They're especially keen to know more about the details of our own planet's history—for example, the importance of methane as an early terrestrial greenhouse gas. But what if a planet were considerably smaller than Earth in a Venus-like location? What if it started out with only half as much water? Or twice as much? Could a planet's size, water content, and levels of carbon dioxide conspire to make it habitable, over long periods, even at the distance of Venus from a Sun-like star? What if a planet were two or three times the mass of Earth but in an orbit even bigger than the orbit of Mars? A massive but remote Earth might be able to hang on to an atmosphere that was rich in hydrogen and other reducing gases throughout its entire lifetime. Alternatively, it might be able to build up an oxygen-rich atmosphere. Can the bio-geophysical parameters be adjusted to produce a habitable rocky world even if, for example, it's twice as far out as Earth is from the Sun?

These are among the fascinating, complex issues that planetary astronomers and astrobiologists are wrestling with at the present time. They're pushing on the limits of the habitable zone at both margins, trying to model the variety of circumstances under which biology is conceivable. At the same time, they have to be mindful that HZs are not static affairs. Just as their boundaries vary depending on the type of star and the details of the planetary system, so a given HZ will shift position over time. The main reason for this is that stars grow brighter as they age. The Sun, for example, had only 70 percent of its present luminosity four billion years ago and eventually (before it evolves to become a red giant) it will double its original brightness. Stellar brightening causes the HZ to migrate outward over time. Consequently, when it comes to long-term habitability, what matters is not where the HZ is at any moment in time but the portion of it that can support life over it's central star's entire main-sequence career—the so-called *continuously* habitable zone.

Yet the traditional idea of habitable zones ignores some important possibilities. Most obviously, it doesn't allow for extremely alien forms of life. If it turns out that some things in the universe can live at extraordinarily high or low temperatures, or based on some exotic chemistry, then nothing we can say about "habitability," given our present knowledge, will be relevant. Also, it has to be said, the original concept of a habitable zone is quite dated. It goes back to the late 1950s, when the Chinese-American astronomer Su-Shu Huang made the first detailed analysis of the types of stars that might be capable of nurturing life. That was long before anyone realized the range of seemingly hostile environments in which life can exist even on our own planet, including deep, hot rocks and undersea hydrothermal vents.

The traditional idea of a habitable zone focuses on the availability of liquid water on Earth-like *surface* locations. It doesn't, for instance, contemplate a subterranean ecosystem on Mars or the possibility of life in underground oceans on the moons of giant planets. Jupiter, for example, lies far outside the Sun's HZ, yet astrobiologists today wouldn't be at all surprised to find microorganisms on Europa and on one or two of the other Jovian moons.

The whole concept of habitable zones focuses on life of a very special kind: surface-dwelling, water-dependent life that derives its energy ultimately from the host star. That a world's *internal* supply of heat might be crucial to sustaining certain kinds of organism doesn't enter into the reckoning; nor does the very real possibility of biological forms that are far outside our experience.

Still, these conventional watery environments are a useful starting point. For one thing, the only kind of life that scientists are going to be able to detect over interstellar distances is surface life, because of its effects on a planetary atmosphere. So for remote detection purposes, it makes sense to begin by looking for surface biology.

Bearing this in mind, let's press on, using the idea of habitable zones for the time being, to consider the potential for life around stars more luminous than the Sun. Following the trend we've already seen, the plain Stefan-Boltzmann equation predicts that the habitable zone moves farther out and widens as stellar luminosity increases. Next up in brightness from Sun-like (G-type) stars are the yellow-white F-types and then the brilliant white A-types, of which Sirius, the Dog Star, is the best-known example. Finally, at the high end of the luminosity scale, are the B-types and O-types—super-hot, blue-white giants with the light output of thousands of Suns.

The larger habitable zones of hot, bright stars at first look promising. Perhaps the more luminous the star, the better its chances of supporting life? Unfortunately, that line of reasoning is spoiled by a couple of other factors. Hot stars give off large amounts of very penetrating ultraviolet radiation that would be more likely to destroy life on a nearby world than to nurture it. Furthermore, hot, bright stars burn themselves out fast. An O-type star with 30 times the mass and 10,000 times the energy output of the Sun would exhaust its main energy reserves—large though they are—in a few tens of million of years. This is nowhere near enough time for conventional life to develop.

We can forget about life-as-we-know-it around the searchlight stars— the O- and B-types. In practice, this is no great loss because the bigger and brighter the star, the more rare it is. O-types account for only about 0.00002 percent and B-types about 0.09 percent of all the stars in the galaxy. Scientists also tend to discount the biological potential of the Sirius-like A-types; although, with a billion-year life-span or thereabouts, it isn't out of the question that they might support primitive, radiation-resistant organisms on sufficiently distant worlds. The F-types, however, look much more promising. Their UV output is high, but not unreasonably so, and their central hydrogen reserves are sufficiently long-lasting, at around four billion years, to give life plenty of time to get established.

At the end of the day, astrobiologists have to ask: of all the stars in the galaxy, what kind offers the best prospects for nurturing life? At the lower range of the luminosity scale, the red dwarfs are an unknown quantity. Maybe they can support life, or, maybe problems to do with planet formation, captured rotation, or stellar flares work against them. They certainly can't be ruled out. At the other extreme, the O- and B-types, and perhaps the A-types too, come across as much poorer bets. Which leaves the middle-of-the-road stars—the K-, G- and F-types—as the focus of astrobiological attention.

That may not seem very exciting. After all, we really don't have any idea of the limits of life. Perhaps the brightest stars in space can support life that develops many times faster than anything we've thought about. Perhaps life can evolve on the surface of a pulsar, or inside a black hole, or far away from any star. No one can yet disprove such possibilities. Given our current state of ignorance, life *might* exist almost anywhere and in any form. But astrobiologists are building a science, not an inventory of plots for *Star Trek*. In their professional capacity, with finite resources and limited funds, astrobiologists need to optimize their chances of finding the first examples of extraterrestrial life with all reasonable speed. So they look at the only certainty they have: that life, based on carbon and water, exists on a rocky planet orbiting a G-type star at a distance of 150 million kilometers. This gives them a safe opening gambit when it comes to seeking life beyond the solar system. Their strategy will be to look for life as we know it in an Earthlike location—in the habitable zone of a reasonably Sun-like star.

Of course, there's no use a star being generally favorable toward life unless it actually has planets going around it. The quest for extraterrestrial organisms is thus inevitably bound up with the quest for extrasolar planets. And, in this field, recent developments have been both rapid and spectacular.

☼

The first planets found outside the solar system seemingly had no right to be there. They're in an outlandish place, and astronomers still don't properly understand how it happened. The planets, of which three have been confirmed and a fourth hypothesized, circle a star called PSR 1257+12, at the remote distance of 1,000 light-years in the constellation Virgo. The second and third planets are roughly three times as massive as Earth and move in orbits that are about a third and about a half the size of Earth's. Nothing remarkable so far, you might think. But the name PSR 1257+12 is a clue that something is very odd. PSR is short for "pulsar." These are no ordinary planets and this is no ordinary star.

A pulsar is a rapidly spinning neutron star—effectively, an atomic nucleus as wide as Pittsburgh. Its density doesn't bear thinking about; neither does the notion that it could have orbiting worlds. For a pulsar is the whirling remnant of a giant star that catastrophically erupted in a supernova explosion.

Unusual perhaps, but what's unreasonable about it having planets? If the old giant sun had a planetary system before it blew up, these are surely just the charred remains—burnt-out cinders of worlds orbiting their stellar wreck. Well, there's a problem with the dynamics. When the big star went supernova, it jettisoned most of its mass into space. The gravity pull of the pulsar wouldn't be strong enough to hold on to whatever planets had been there before, given the speed at which they would have been moving in their orbits. It would be like a strong athlete who was wheeling around an Olympic hammer, suddenly finding that he had only the strength of a child. The hammer would instantly fly out of his grasp.

That leaves only one possibility: the pulsar planets must have formed *after* the supernova. Somehow, they must have gathered themselves together from dregs of the gas and dust shot out of the dying star—a pathetic retinue forged not in the wake of stellar birth but of stellar death. The physics remains sketchy. But, to put it mildly, these curious worlds are not high on the alien-hunter's target list. Anything in the vicinity of a pulsar would be continuously strafed by high-energy radiation, including X-rays and gamma rays—the perfect agents of sterilization.

Four years after the discovery of the pulsar planets, however, came more satisfying news for the astrobiologist. Several teams had been racing to be the first to make an announcement. In the end, the honors went to the Swiss pair of Michel Mayor and Didier Queloz at the Geneva Observatory. They had found a planet in orbit around 51 Pegasi, a G-type star like the Sun, some 50 light-years away. This first extrasolar world of a normal star had about half the mass of Jupiter, putting it in the gas giant category. But its orbit was fantastically small. It averaged only eight million kilometers in radius—one-seventh the distance of Mercury from the Sun!

Nothing that astronomers had ever seen or theorized prepared them for this. They'd expected to find other Jupiter-like planets, of course. But they'd also expected to find them in wide orbits, similar to those of the gas giants in our solar system. The new-found world of 51 Pegasi was racing around its star in an orbit 20 times smaller than the Earth's and in a "year" (the time to complete one orbit) that lasted barely four days. How was that possible?

Perhaps it was a fluke. But no, news soon came of another, similar discovery. At San Francisco State University, Geoff Marcy and Paul Butler announced they'd detected a planet around Tau Bootis, an F-type star, somewhat hotter and brighter than the Sun. This planet was even more extreme than the one in Pegasus. It boasted over three times the mass of Jupiter, yet orbited only seven million kilometers from its star.

At first, people wondered whether a gas giant could conceivably hold on to its hydrogen, helium, and other volatile materials in such a torrid location. Instead of a cloud-top temperature of –150° to –220°C, as in the case of the gas giants we're familiar with, the "hot Jupiter" of Tau Bootis never cools below about 1,700°C. There was even speculation that it might not be a gas giant at all, but a monstrous cosmic cannonball made of solid metal. Yet where could so much metal come from? The giant cannonball idea really wasn't feasible. In any case, calculations soon showed that the powerful gravitational embrace of a gas giant *would*, against intuition, enable it to retain its gassy bulk even at a temperature high enough to melt iron.

Granted that a gas giant could survive in the furnace-like interior of a planetary system, how could it possibly have *formed* there? The simple answer is that it couldn't. These big worlds in sub–Mercurian orbits almost certainly formed much farther out, at distances like those of Jupiter and Saturn from the Sun. Something happened that dramatically altered their course. But what? Within hours of the announcement of the world around 51 Pegasi, astronomers knew that their preconceptions had been wrong— that planetary systems were going to prove to be much more diverse and subject to violent rearrangement than anyone had anticipated.

That suspicion was quickly backed up by a second discovery by Marcy and Butler early in 1996. It was an extrasolar planet, a companion to the Sun-like star 70 Virginis, in a very different kind of unusual orbit. Whereas the worlds of 51 Pegasi and Tau Bootis were of the "hot Jupiter" persuasion, that of 70 Virginis belonged to a new and again unforeseen category. Its orbit was extremely elongated—much more so than that of any planet in the solar system, including Pluto. There was some excited talk, early on, that the planet of 70 Virginis might be able to support liquid water, because it passed through the habitable zone of its host star during its annual circuit. But any such prospect vanished when the details were looked at closely. First, the orbit is too small; most of it doesn't lie in the habitable zone at all. If it were in the solar system, it would come much nearer to the Sun than Mercury at one extreme and not much further away than Venus at the other. Furthermore, this strangely moving planet, far from being reasonably Earth-like has about *seven* times the mass of Jupiter.

What was such a colossus—or any planet—doing on such a strange path? The architecture of our own solar system had encouraged astronomers to suppose that planets everywhere follow more-or-less circular orbits, ranging in size from a few tens of millions to a few billion kilometers across. Yet, no sooner had we begun finding other solar systems, than we turned up two

Jupiter-like worlds virtually hugging their central stars, and a super Jupiter in an orbit as oval as an egg.

Over the past few years, several dozen more extrasolar planets have been found around Sun-like stars. Most are bigger than Jupiter, some are smaller than Jupiter but bigger than Saturn, and a handful are roughly Saturn-sized. There can't be any doubt that they are gas giants. Most continue to fall into the remarkable new categories of hot Jupiters and "eccentric" Jupiters— eccentric in the mathematical sense of following a path that is a long, drawn-out ellipse. Few of these worlds have orbits that astronomers a decade ago would have thought likely or even possible. What does it mean?

An answer began to emerge when astronomers went back to thinking about how planetary systems are made. All planets must start off in fairly circular orbits, for the simple reason that the material out of which they form is part of a great circular, rotating disk. In addition, all planets that are a lot bigger than Earth must take shape at distances similar to those of our own gas giants, because only in these regions is the temperature low enough to let them accumulate their enormously thick atmospheres of light gases.

The initial arrangement, then, of all planetary systems probably follows a standard pattern, with smaller, rocky worlds occupying the inner, warm zone and larger, gassy worlds located further out, all of them in orbits not far from perfect circles. It so happens that our solar system has retained this overall design. But, even within the Sun's well-ordered domain, there are signs of unruly behavior billions of years ago. As we've seen, the Earth may have been hit by a Mars-sized planet which led to the creation of the Moon. Uranus, too, evidently took a mighty blow in its infancy, hard enough to tip it over so that it now spins on its side. Recent theoretical work suggests that the orbits of Saturn, Uranus and Neptune have all grown since the earliest days of the solar system. In particular, Neptune may have moved out by more than a billion kilometers, pushing Pluto outward as well, and shifting it from a nearly circular orbit to one that is markedly oval. More spectacularly, in these early dynamic times, untold numbers of smaller objects—simply because they *were* smaller—were tossed around like paper bags in a hurricane.

Comets are a prime example. Having assumed circular orbits when the solar system first began to form, they quickly found themselves scattered this way and that by the gravitational slingshot effect of the new-born gas giants. One possible result of a comet's passage close by a massive planet is that it's thrown onto a highly elliptical orbit that takes it alternately near to the Sun and then much further away. Exactly the same effect, astronomers believe, explains the remarkably elongated paths of the eccentric Jupiters found around other stars. Computer simulations show that if two massive planets form in orbits that are fairly close together—up to a few Earth–Sun distances apart—their mutual gravitational tug will hurl both off course. One planet will swerve inward onto a much smaller and eccentric orbit like that of the giant around 70 Virginis; the other will be slung outward onto an oval circuit that carries it much further from its central star.

In an extreme case, an inward-thrown planet may even be vaporized and destroyed. This could explain a surprising fact that a team of astronomers at the European Southern Observatory uncovered in 1999 about a star known as S50. When the astronomers measured the concentration of elements at the star's surface, they found that the light metal lithium was 100 to 1,000 times more plentiful than expected. Because S50 showed no other unusual features, they concluded that they'd come across a bad case of stellar dyspepsia brought on by swallowing a gas giant. At some point in S50's two-billion-year history, a couple of big planets in orbit around the star may have experienced a close encounter, sending one of the pair on a direct collision course with its sun.

In other cases, where two big planets toss each other around like zealous dancing partners, one of them might narrowly avoid careening into the central star. If its inward-plunging orbit carried it very close, then it might gradually be drawn by the stellar gravitational field into a more and more circular path of small radius, until it became a hot Jupiter like 51 Pegasi.

Theorists know of a couple of other ways—which may actually be more common—by which gas giants can become hot Jupiters. These involve interactions at an early stage between protoplanets (planets in the making) and the protoplanetary disk in which they're embedded. One possibility is that a protoplanet gets gravitationally tied to gas in the disk that is spiraling in toward the still-forming star at the center, only to break free at the last moment and avoid the fatal swan-dive. In another scenario, a protoplanet becomes progressively drained of orbital momentum through numerous sling-shot and other small gravitational interactions with planetesimals and comets, until its orbit shrinks all the way down to just a few million kilometers across.

✷

The early years of extrasolar planet detection have made one thing clear. Young planetary systems are the scene of enormously violent and chaotic events that can hurl worlds around and change their orbits dramatically.

For astrobiology, the most encouraging news is that planets in general seem to be exceptionally common. Not only are astronomers adding, almost monthly, to the list of known extrasolar worlds, but they've also peered into stellar nurseries, such as the Orion Nebula, and found that a high proportion of young stars are surrounded by dusty disks of the kind in which planets form.

On the other hand, the thought of worlds being tossed around on a routine basis is a bit disconcerting. If something the size of Jupiter, or bigger, gets diverted into the inner part of a young planetary system, any smaller and potentially life-bearing worlds already there are likely to be severely disrupted. Earthlike planets could be torn from their orbital moorings and cast into the central sun or thrown clear into interstellar space. Many billions of dark Earths could be roaming the galaxy, cut adrift from the star systems in which they were made.

Earth-sized planets could also form much farther from the Sun, in the cold outer regions with the gas giants. Caltech's David Stevenson, whom we met earlier in connection with plate tectonics, has asked: What if there had originally been five or ten such worlds in orbits that crossed Jupiter's? Because of their low temperature, they would have been able to accumulate and hold on to thick atmospheres of molecular hydrogen. But, one by one, as they strayed too close to Jupiter, they would have been thrown in toward the Sun, captured as moons, or slung away from the Sun altogether. Then what? In the case of the solar system escapees, Stevenson has made the intriguing suggestion that interstellar planets like these—Earth-sized and rocky but swaddled in dense, hydrogen-rich atmospheres—might just be capable of supporting life. Although deprived of the warmth and light provided by a home star, they would have a healthy internal supply of heat, just as the Earth does. What's more, the thick hydrogen blanket would be extremely effective in preventing this geothermal outflow from leaking away into space. Instead, it might keep the surface temperature high enough for liquid water to exist and primitive organisms to evolve.

The possibility of another realm of life has been opened up by the suggestion of sub-surface oceans on Europa and other large moons in the solar system. Many of the newly found gas giants circling other stars also presumably have big moons subject to tidal heating. If the theory of Europan water and life holds up, biology in this type of environment could be extremely widespread, though perhaps also quite rudimentary. Astrobiologists are especially keen, of course, to know how often terrestrial worlds form and survive, reasonably unscathed, within the habitable zones of their stars. This is still the only situation in which we know for certain that life has come about. Yet the roll-call of extrasolar planets discovered so far seems to present a discouraging picture. Almost all of them are gas giants in weird orbits. Hot or eccentric Jupiters seem to be the order of the day. Does this mean planetary systems similar to our own, and Earthlike worlds in particular, are an oddity? Is astrobiology doomed to be dull, and should SETI researchers start looking for other jobs?

6

Rare Earths
and Hidden Agendas

While most astrobiologists are upbeat about finding all kinds of life in the universe, a minority have their doubts. According to this skeptical school, our planet is so special that all but the simplest of organisms are likely to be uncommon elsewhere. Although many of the arguments on both sides are new, the debate about the uniqueness of Earth is as old as civilization. What's more, just as the debate in earlier times was by no means purely scientific, neither is it today.

☼

The idea that our planet may be biologically almost unique was put under the spotlight by *Rare Earth: Why Complex Life is Uncommon in the Universe*, a book by Peter Ward and Donald Brownlee, a paleontologist and astronomer, respectively, at the University of Washington in Seattle. First published in January 2000, *Rare Earth* is a polemic for the view that whereas microbial life is likely to be widespread on other worlds, multicellular organisms—and intelligent life in particular—will prove to be scarce. The book has sold well, attracted an unusual amount of media attention, and has provoked comment and controversy among scientists and non-scientists alike. As its reviewer in the *New York Times* pointed out, "*Rare Earth* . . . is producing whoops of criticism and praise, with some detractors saying that the authors have made their own simplistic assumptions about the adaptability of life forms while others call it 'brilliant' and 'courageous.'" *The Times* of London wrote, "If they are right it could be time to reverse a process that has been going on since Copernicus."

Many on the Religious Right have embraced the book as a vindication of their belief in the special nature of the Earth, while SETI researchers, said the *New York Times*, "See the book as a heretical assault that could endanger the financing of the hunt." Ward feels that "An awful lot of astrobiologists, NASA itself, are threatened by this"—an irony, considering that his co-author is chief scientist with NASA's *Stardust* mission to capture samples of interplanetary and interstellar dust. "Somehow, I don't think we really appreciated that this was going to raise such hackles," said Brownlee.

Reaction to the book among scientists has been mixed. Geoff Marcy, a leading seeker of extrasolar planets, suspected that *Rare Earth* might spark a revolution in thinking about extraterrestrial life. Generally, however, the scientific response has been cooler. In his review of the book for *Science*, NASA Ames astrobiologist Chris McKay describes the authors as "[making] the case (if not always convincingly) that the situation on our Earth is optimal for the development of complex life." He adds, "We have only one example of life" and the "assessment of [the] probability" for the development of life "is uncertain at best." A similar point is made by the physicist Lawrence Krauss of Case Western Reserve University, writing in *Physics Today*:

> Ward and Brownlee summarize clearly the developments over the past few decades that reveal the complexity of the evolution of advanced life on Earth. However, demonstrating the complexity of a process is different from demonstrating that the end result is rare . . . It is undeniable that the specific route that led to modern terrestrial life-forms is remarkably complex and probably has a small absolute probability. But the same can be said for the series of events that led me to my computer this evening.

There's no doubt that the authors have done a service by challenging conventional wisdom; strong opposition is always good for the parliament of science. But how valid are their arguments? Ward and Brownlee are the first to admit they're not the architects or originators of what they call the Rare Earth Hypothesis. As Ward explains, "I did not go into the debate prior to writing our book. It just seemed intuitive." Most of the issues they raise stem from the work of other scientists over the past decade. These issues, in turn, are the latest contributions to an idea that stretches back more than 2,000 years.

☀

The Rare Earth controversy has its roots in ancient Greece, where philosophers asked: Are there other worlds like ours harboring other life like us? Of course, the Greeks' notion of the cosmos was entirely different from ours. The Hellenic *kosmos*, in its most widely accepted form, placed the Earth at the center of a series of concentric revolving spheres to which the Moon, Sun, planets and stars were fixed like little lights. There was no conception of stars as huge balls of hot gas, or of orbiting extrasolar worlds. This single-world geocentric *kosmos*, in which mankind found itself at the focus, was the only one of which we could be directly aware. To Aristotle, Plato and their followers, it was the only one possible, because more than one *kosmoi* didn't mesh logically with their other beliefs.

The rival Greek school of atomism, however, disagreed. In this vision of nature, all things originated through the chance coming together of little bits of matter in endless combinations in an eternal, infinite void. Worlds and beings of every description were actually predicted, as Epicurus explained in his *Letter to Herodotus*, "[T]here are infinite worlds, both like and unlike this world of ours . . . we must believe that in all worlds there are living creatures and plants and other things we see in this world."

Yet these "infinite worlds" with their livings things—these other *kosmoi*—weren't accessible in any way. Atomists didn't think of them as being somewhere that one might, even in principle, travel to, like the planets of other stars. Instead they were separate and self-contained universes, each with an inhabited Earth at its heart.

More than a thousand years later, Aristotle's scheme became the cosmology of choice in Europe—approved by the Church of Rome and woven into medieval Christian teachings. A single inhabited world sat well with the doctrine of incarnation and redemption, but multiple Earths and multiple sentient races did not. For the inhabitants of these worlds to receive salvation, Jesus would have to be born and sacrificed on every one of them. Just as seriously, the atomist cosmos called into question the unique relationship between God and Man that, according to some interpretations, the Bible implied.

As long as the Earth was the physical hub of creation, it was easy to believe it was unique. But when Copernicus began the transformation of our cosmic perspective by putting the Sun at the center of the solar system, suddenly our planet began to seem much less privileged. Five hundred years later, the full extent of our mediocrity has become startlingly clear. The Sun is just another star, one of many billions, within one of many billions of galaxies. And the Earth, it seems more and more evident, is one among a host of planets far outnumbering all the grains of sand on all the beaches of the world. With the rise of modern cosmology, the Copernican Revolution, in the physical sense, is essentially over.

Now attention has shifted to Earth's *biological* status. Is that also ordinary? Is the Copernican Revolution about to transform the life sciences as it has astronomy and physics? The majority of astrobiologists think so. Virtually all those engaged in SETI—the search for extraterrestrial intelligence—believe so, too. But, in the background, is a protesting murmur that has risen and fallen in waves over the past century.

It was loudest in the 1920s and 1930s, when two factors conspired to make terrestrial life seem exceptional. First was the opinion, widespread among biologists at the time, that the steps leading to life were highly improbable. Second was the theory, in vogue with astronomers during the same era, that the planets of the solar system had formed in the wake of a near-collision between the Sun and another star. If planetary systems came about only as a result of such incredibly unlikely encounters, they, and life, were bound to be unusual. By the 1950s, the consensus had shifted to life's origins being easier and planets common.

Then, in the late 1970s, a new Rare Earth argument surfaced. It came during a period of gloom for those who would later be called astrobiologists, following the failure of *Viking* to find life on Mars. Some astronomers, led by Michael Hart, suggested that the habitable zone around a star—the region in which an Earth-like planet would be able to support liquid water—was much narrower than had been thought. According to Hart's calculations, if Earth had been just one percent farther from the Sun it would have become permanently encased in ice. But, a few years later, the pendulum swung back again, when James Kasting and his colleagues at Penn State pointed out the importance of extra greenhouse heating if the Earth's distance had been somewhat greater. More recently, as we've seen, the outer boundary of the Sun's HZ has been pushed back further—to the fringes of the asteroid belt—by the discovery that dry ice clouds provide an additional warming mechanism.

During the 1990s, fresh claims were made, implying that our world and its advanced life might be exceptional. These arguments, advocated by a number of scientists, have been drawn together as a multi-pronged thesis in Ward and Brownlee's book. The Rare Earth Hypothesis doesn't claim that life of every sort is rare. On the contrary, it's orthodox in pointing to the early appearance of terrestrial organisms and the tenacity and hardiness of extremophiles as compelling evidence that microbial life might arise routinely elsewhere in space. The controversy stems from its suggestion that *complex* life, in the form, for example, of animals, is scarce across the cosmos. In order for life to evolve and survive beyond the microbial stage, Ward and Brownlee argue, a very special combination of factors must prevail. These factors are said to be both rare in themselves and absolutely indispensable to complex life; therefore, according to the Rare Earth Hypothesis, a planet supporting such life must satisfy *all* of these criteria:

- It must have a large, nearby moon. But, since such moons probably form from chance collisions between planet-sized objects, they're presumed to be unusual.
- It must experience the right (moderate) level and timing of catastrophic events to promote biological diversity without extinguishing life. Yet the odds of a near-repeat of Earth's history are very small.
- It must be an Earth-like world in its star's continuously habitable zone. But most of the extrasolar planets found to date are giant planets in unusual orbits that would have hurled Earth-like worlds out of their stars' HZs.
- The planetary system must contain a Jupiter-like world in a Jupiter-like orbit to protect the inner worlds from being overly bombarded. None of the new-found worlds fits this bill.
- The planet must orbit a solitary, stable, Sun-like star with a relatively high heavy element content, or "metallicity." Stars of such composition make up only a small fraction of the total stellar population.
- The planet must have ongoing plate tectonics. This requires that the planet have both liquid water and sufficient mass to have a large internal store of heat.
- The planetary system must move within what has been called the "Galactic Habitable Zone." This is a relatively narrow belt that excludes most of the stars in our galaxy.

The modern debate over Rare Earth, on one level, is a debate over how well these claims hold up under close scrutiny. As we shall see, this is not the only level of debate, but it's worth examining each point in turn.

A Large, Nearby Moon

In 1993, the French astronomer Jacques Laskar published the results of a study on habitability that focused on planetary dynamics, and in particular the way planets wobble in their orbits. He concluded that a planet's ability to support life might depend crucially on the presence of a large moon.

Any planet that spins develops a slight bulge at its equator; if its axis is tilted, the tug of the central star on this bulge causes the planet's axis to wander slowly around like a wobbling top—an effect known as precession. In addition, there's an orbital precession driven mainly by the gravitational pulls of giant outer planets. Imagine the Earth's orbital plane to be a dinner plate, with the Sun a mound of mashed potatoes in the middle and the Earth a lump of chewing gum stuck to the rim. The incessant tugging of Jupiter and Saturn, acting against the Sun's stronger pull, makes the plate—the plane of Earth's orbit—wobble as it turns.

On their own, orbital wobbles aren't significant. However, the orbital and axial precessions of a planet may interact with surprisingly dramatic results. This happens if a planet's axial wobbling is slow enough to fall into rhythm—to resonate—with the torpid rocking motion of its orbit. Then the central star's tidal pull on the planet's equatorial bulge causes the spin axis to keel over at a greater and greater angle before gradually returning to a more upright position.

Since 1973, it's been known that the tilt of the Martian axis (currently 25 degrees) rocks back and forth between 15 and 35 degrees over a period of several hundred thousand years. But Laskar found that over tens of millions of years, the swing is bigger—between 0 and 60 degrees. His calculations showed that Earth's axial tilt would slew even more wildly—between 0 and 85 degrees—were it not for one saving factor: the Moon. The lunar gravitational field has sped up the rate at which the Earth's axis wobbles so that it never falls into resonance with the slow orbital wobbling. This was crucial, Laskar argued, because a planet that alternated between spinning upright and on its side would experience such huge climate swings that it might make life impossible.

Guillermo Gonzalez, an astronomer at the University of Washington and a close associate of Ward and Brownlee's, has even argued that there is a connection between life and the occurrence of solar eclipses. Unless the Moon were as big and far away from us as it is—making it appear as large as the Sun in the sky—intelligent life wouldn't have been able to evolve here.

But it's now clear that Laskar's findings rule out neither simple nor complex biology. Contrary to Ward and Brownlee's assertion, big moons may not be rare, necessary, or even desirable to the emergence of higher forms of life. Recent computer simulations by Eugenio Rivera of NASA Ames and his colleagues suggest that small planets with big moons are likely to be quite common. As many as one in three Earth-like planets in their infancy may be struck hard enough by other large objects to make big moons, and one in twelve struck at a time when its tilt is sufficiently mild for it to be stabilized at a terrestrial angle (currently 23½ degrees) or less.

On the question of whether a big moon is crucial to the emergence of complex life, there are two points. First, if the Earth had been deprived of a big moon then, as Seth Shostak of the SETI Institute points out:

> Our planet would spin faster—fast enough, in fact, to stabilize it against major tipping.* In addition, even if an Earth-like planet occasionally does spin flip, it will spend 10 million years or more doing so. Life can probably adapt to such slow changes. Indeed, it already has, during episodes of polar wander on Earth.

Second, new biological possibilities are opened up for planets that *do* periodically roll on their sides for want of a stabilizing satellite. In 1997, James Kasting and Darren Williams at Penn State calculated what our climate would be like if the Earth were tipped on its side (as Uranus is) and located 1.4 times further from the Sun, at a distance of 210 million kilometers. They found that, given the extra greenhouse heating due to increased levels of carbon dioxide, conditions would be positively balmy. The equator would be at a steady 11°C, whereas the poles would never rise above 46°C or fall below 3°C. There'd be no ice anywhere, except on the top of tall mountains.

Finally, there are circumstances in which having a big moon would be detrimental to life. As Alan Boss of the Carnegie Institution explains:

> It is true that, in the present solar system, the Moon does act to stabilize the spin axis of the Earth. And it does that by accelerating the precession of the Earth's spin axis, making it precess much faster than the orbit of the Earth does. But in another planetary system where the planets might be spaced more closely, you might have a giant planet closer to the terrestrial planet. The orbit of the Earth-like planet would precess much faster and actually, the presence of a moon would be damaging to the stability of the spin axis. Thus we really need to know quite a bit about a planetary system as well as the types of satellites which may be around these planets to know whether the spin axes are stable or not.

---

*Higher rotational speeds would, however, result in stronger surface winds.

"Is the lack of a large moon sufficient to prevent microbial life from evolving into animal life?" ask Ward and Brownlee. "We have no information," they candidly admit. It's a problem that pervades the Rare Earth Hypothesis. We've no way of knowing how different environmental histories or circumstances affect the trajectory of biological development, because we have no comparative data from other planetary systems. And, faced with so many variables and uncertainties, current theoretical models simply can't fill the void.

Catastrophic Impacts

This problem becomes more conspicuous when we look at the next item on the Rare Earth laundry list: the evolutionary role played by catastrophes, such as major collisions. Impacts by comets and asteroids, it's generally agreed, can have both positive and negative effects on a planet's biosphere. As Carl Sagan said, "Comets giveth, and comets taketh away." Quite apart from any role they play in delivering organic matter and water, impacting objects may both benefit and harm the prospects for biological evolution. If they battered a world often and hard enough, they might eradicate life or prevent it ever from crawling out of its deep, dark hiding places. But planetary traumas can also act as stimuli. Like wars, they cause horrifying devastation and loss of life but also lead to accelerated development. Rocketry, aviation, electronic computers and radar all progressed rapidly during the Second World War. In a similar way, natural planetary catastrophes can drive biological evolution to new heights.

Think about the disaster that happened 65 million years ago. An asteroid some ten kilometers wide smashed into what is now Mexico's Yucatan peninsula, causing such havoc and short-term climate upheaval that it annihilated 75 percent of the planet's life-forms. Ecologies were massively disrupted; the dinosaurs were annihilated along with many other creatures, and mammals and birds stepped up to fill the job openings suddenly made available by the departure of their scary reptilian superiors. We'll never know what might have happened if Class Dinosauria had lived on. Perhaps some of the "terrible lizards" would have evolved into smart, mobile-phone-using lizards. What's clear is that, without the Cretaceous–Tertiary boundary event, humans wouldn't be here. From our perspective, and perhaps from the point of view of the accelerated rise of intelligence in general, the K.T. calamity proved to be good news.

Global traumas in earlier times, too, appear to have thrown evolution into high gear after long periods of relative torpor. Most spectacularly, the rise of animal life has been linked to one of the biggest chill-outs in Earth's history. Evidence is mounting that our world was almost completely encased in ice several times between about 750 million and 580 million years ago. If the "Snowball Earth" theory, stemming from the work of W. Brian Harland at the University of Cambridge in 1964, Joseph Kirschvink at Caltech in 1992, and most recently by Paul Hoffman at Harvard, is right, the global deep-freezes were so extreme they wiped out most of the life on the planet, turned the top kilometer of the oceans to solid ice, and plunged surface temperatures everywhere to between –20°C and –50°C.

The timing of Snowball Earth is intriguing because before about 600 million years ago life on Earth consisted mostly of microbes. For three billion years or more, hardly any individual thing alive was big enough to see. Then suddenly (geologically speaking) there was a proliferation of new life—the so-called Cambrian explosion—when, in an unprecedented burst of natural creativity, sponges and jellyfish, sea anemones and coral polyps, and soon after these, worms, mollusks, creatures with jointed legs, and finally animals with backbones appeared on the scene. All of this burgeoning of elaborate many-celled forms began after the final thawing of the Snowball and the return to moderate conditions.

A much earlier Snowball event, from 2.5 to 2.2 billion years ago, has been implicated by Kirschvink in the other great biological revolution on this planet—the emergence of eukaryotic cells. These, in turn, paved the way for more sophisticated life-forms, including higher plants and animals.

Life began during a time of trauma, some four billion years ago. And according to a recent study, much of the period over which complex organisms have developed to their present stage—the past 400 million years or so—has been a busy one for objects colliding with the Earth. This study involved measuring the age of tiny glass beads, known as cosmic spherules, in samples of lunar rock brought back by Apollo 14 in 1971. Spherules are formed when meteorites crash into the moon and melt rock at the point of impact. Researchers from the University of California at Berkeley and two other institutions, using a precise isotope method, dated hundreds of them in soil recovered from Mare Imbrium (the Sea of Rains). They were found to come from 146 different craters in the Imbrium region, and their age distribution spoke of an era of intense bombardment over the past few hundred million years. Assuming the Earth suffered a similar level of bombardment, it seems that life may have taken its final steps to the advanced state we see around us today during a time of impacts that was hardly less intense than the primordial bombardment phase during which life originated.

Finally, humanity itself evolved from ape-like ancestors during a period of moderate global cooling and drying a few million years ago. The final cultural push, which took anatomically modern man to the brink of civilization, happened in the midst of the ice ages that gripped much of Europe over the past few tens of thousands of years. Advanced life and civilization are here courtesy of disaster, devastation, and worldwide freeze-ups.

What with one thing and another, we've ample evidence that major stresses to the biosphere can have positive long-term effects on evolution (as seen from an advanced creature's point of view!) The question is: what level of stressing is *optimal* for the development of complex life? It's clearly a question of balance. If disasters happen too often, complexity could be repeatedly destroyed before it can establish itself. If there were a complete absence of big planetary incidents, there mightn't be enough stimuli to promote evolution. Somewhere between these extremes must lie just the right level of environmental stress to push life forward at its maximum possible rate.

What is this optimal level? How often, and when, should a planet experience a cathartic event so that it will incubate complex life in the most effective way possible? The answer is, we've absolutely no idea—again, because we have no comparative data from other evolutionary histories. Yet despite our ignorance, the Rare Earth Hypothesis is ready with an answer: the very planet on which we live experienced the ideal level of stimulation to make advanced life possible. How lucky it was! Even when Earth did get hit, it was just hard enough and at just the right time to encourage the survivors of the impact to adapt and exploit whatever new opportunities the disaster opened up. When it froze over, it was exactly at those opportune times when it would most effectively promote the rise of complexity. Our very own planet is just about as good as it gets from the point of view of allowing life to evolve as fast and as far as it can.

Such arguments, constructed after the fact and based on a solitary example, set the alarm bells of skepticism ringing. As Athena Andreadis, a neurologist at Harvard Medical School, points out:

> In science, theories cannot be identical to their prediction, nor can that prediction be trivial. In fact, the Rare Earth theory is neither hypothesis nor prediction, but a description of how life arose on Earth . . . Their [Ward and Brownlee's] oft-repeated statement that both Earth and humans are unique is neither novel nor contested—nor helpful in predicting life elsewhere.

But Ward and Brownlee go further: they actually pick and choose the factors that best suit their case. Without the Moon, they claim, the emergence of complex organisms would have been frustrated by large climate swings. Yet they go on to endorse the idea that what may have been the biggest climate variations in our planet's history—from Snowball Earth to subsequent warming—were key catalysts for the rise of multicellular life. They even point to a possible further trigger for complex life in the form of a so-called inertial interchange event (IIE) which, in 1997, Kirschvink and his colleagues David Evans and Robert Ripperdan suggested may have taken place over a 10- to 15-million-year-period at the time of the Cambrian Explosion. According to the IIE theory, the Earth's spin axis underwent a 90-degree change in direction relative to the continents. This change, speculate Ward and Brownlee, "would have fragmented ecosystems and could have prompted evolutionary diversification." Four chapters later, however, Ward and Brownlee do a spin flip of their own, arguing that "If the polar tilt axis had [in the absence of the Moon's stabilizing influence] undergone wide deviations from its present value, Earth's climate would have been much less hospitable for the evolution of higher life forms." Which is it to be? We're not in any position to say what sort of climate disturbances tend to hold evolution in check and which, in the long run, spur it on. Yet, time and again, this is the game that *Rare Earth* tries to play.

At this point we can begin to glimpse the frailty of the Rare Earth position. Its problems stem from an unsubstantiated initial assumption—namely, that complex life (as distinct from primitive microbial life) is difficult. How do we know that? Complex life is difficult, say Rare Earth supporters, because it took so long—almost four billion years—to evolve on Earth. If it's difficult, then in order to have nurtured it Earth must be special. And then the laundry list starts: Our world's a special size with a special kind of moon. It goes around a certain kind of star, at a very specific distance, in a very particular kind of solar system, and has a unique history. Its climate, record of impacts—everything was just right on this precious bit of real estate for complex life to emerge. What are the odds of finding *all* these factors together elsewhere? Very low. Therefore, complex life is rare.

But hold on. Just because it took complex life four billion years to evolve here, doesn't mean it *has* to take that long. We haven't a clue whether or not the Earth was ideal for raising complex life. We *do* have pretty good evidence that global upheavals of various kinds are generally good for encouraging life to get rid of the dead wood, experiment, adapt and explore new survival strategies. We do have evidence to suggest there were long periods of Earth's history where very little biological progress was made. *Maybe*, worlds to which stressful things happen more often, within reason, arrive at their equivalent of backbones, brains and biologists a lot faster. Maybe if the Earth had suffered more trauma during its first few billion years it would have made a much speedier transition from single-celled to many-celled organisms. Or, maybe some other aspect of the Earth held back the jump to complexity.

NASA Ames astrobiologist Chris McKay thinks that not only complex life but human-level intelligence *could* evolve, from scratch, in only 100 million years—one-fortieth of the time it took here. Plate tectonics on Earth, he suggests, may have greatly delayed the appearance of an oxygen-rich atmosphere. Ward and Brownlee counter by saying that only plate tectonics could maintain a stable oxygenated habitat over *billions* of years—which may be true. But it's irrelevant. Once an intelligent, technological species has evolved, it will (judging by our own rate of development) be space-faring and star-colonizing within a period that's completely insignificant on a geological scale. The fact that its world of origin might be doomed wouldn't matter: there are billions of other planets out there to settle and call home.

Alternatively, it may be that complex life does take a long time to mature. Perhaps four billion years or so is the cosmic norm. However, that doesn't mean it can only come about through an exact repeat of the circumstances and incidents that happened on Earth. If you start off by believing it couldn't have happened in a shorter time—perhaps *much* shorter—or by a different route, you've already implicitly assumed there's something unusual about our world. You've assumed that factors like cosmic collisions, climate history and location were fine-tuned here for the development of complex life. You've accepted the truth of the Rare Earth position before you've even starting presenting arguments to defend it.

The currently unjustifiable basic premise of the Rare Earth Hypothesis—that complex life is hard to get off the ground—leads its advocates into the trap of going out of their way to find reasons why Earth is special. This passage is typical of the extent that Ward and Brownlee are prepared to go:

> *If* the Cambrian Explosion was necessary for animals to become so diverse
> on this planet, and *if* the inertial interchange event occurred as postulated,
> and *if* the Cambrian IIE event contributed to the Cambrian Explosion or
> even somehow was required for the Cambrian Explosion to take place, then
> Earth as a habitat for diverse animal life is rare indeed.

Having got it into their heads that the Earth is special, every terrestrial idiosyncrasy they come across becomes a point in favor of their position. But what matters is not whether there's anything unusual about the Earth—there's going to be something idiosyncratic about *every* planet in space. What matters is whether any of Earth's circumstances are not only unusual but also essential for complex life. So far we've seen nothing to suggest there is.

Extrasolar Planets

As we saw earlier, the first planets to be found around Sun-like stars came as a shock to astronomers. Jupiter-scale worlds in tiny or very elliptical orbits hadn't been anticipated. If these were truly representative of the bulk of planetary systems in the galaxy, the ambitions of astrobiologists would be seriously dented. Jumbo planets yanked out of their original orbits into such radically different trajectories are bound to have scattered any Earth-like planets that once basked in their stars' habitable zones.

But before we rush to conclude that our solar system must be unusual, it's important to understand the huge selection effect at work in these early stages of planet detection. Imagine you're a visiting alien who decides to study the Earth with an exceptionally powerful telescope set up on the Moon. The telescope will show detail as small as two and a half meters across, and you start by trying to detect terrestrial animal life. What do you see? Elephants, rhinos, water buffalo, and so forth, moving about on the African savanna; polar bears in the Arctic; camels—just visible—in the Sahara. Should you infer from your observations that all animals on Earth are at least two and a half meters long, and that the oceans, jungles and forests, into which your telescope can't penetrate, are devoid of animal life? Hardly. You recognize the limitations of your instrument and survey method, and realize that there are almost certainly other living things on Earth that can't be resolved by your equipment or are hidden from view. You realize, too, that the way of the universe is for little things to be a lot more common than big things. If you can see things as small as you're capable of seeing, it's a safe bet that there are many more even smaller things you can't see.

Exactly the same is true of extrasolar planets. The detection method that's proved most successful to date depends on a planet's gravitational effect on its central star. Naturally, the effect on the star is incredibly small because planets are much less massive than the stars they orbit. Even so, with sufficiently sensitive instruments the pull can be measured. A planet causes its star to wobble slightly, back and forth in space, as it orbits around and draws it first one way and then the other. If these wobbles occur at least partly along the line of sight from Earth, they can be detected. The best way to do this is to look at characteristic dark lines—absorption lines—crossing the spectrum of light from a star. As the star is pulled a little toward the Earth by its unseen companion, these lines are shifted slightly toward the blue end of the spectrum; as the star is tugged the other way, the lines move toward the red. By measuring the amount by which the lines are displaced from their normal positions and the time taken for one complete cycle of back-and-forth movement, researchers can obtain a minimum value for the mass of the planet (a point we'll come back to in a moment), and an accurate picture of the size and shape of its orbit.

The biggest stellar wobbles, and therefore the easiest to detect, are caused by massive planets in small orbits. That's why, in surveys of nearby Sun-like stars, astronomers have started out by finding lots of hot Jupiters and eccentric Jupiters. It's also why they've tended to find single planets rather than entire planetary systems. Extraterrestrial astronomers using equally sensitive methods to study our own solar system across tens of light-years might only be able to detect Jupiter, and even then the wobbling they measured would be a lot less than that caused by most of the extrasolar planets found so far.

Extrasolar planet searches have tended to focus on Sun-like stars out to a distance of 100 to 150 light-years. Around *six percent* of these stars, they've found hot or eccentric Jupiters. This is purely a selection effect. Given that such worlds exist, they're bound to be the ones we find first. Ironically, if these unexpected worlds hadn't existed, very few extrasolar planets would have been found over the past few years of searching and Rare Earth punters might be using *that* result as one of their arguments.

Current search methods (with one exception, mentioned below) aren't sensitive enough to pick up planets much lighter than Saturn. They're certainly incapable of registering anything as small as Earth. The other crucial point is that reliable detection takes *two* full orbits—one for the initial alert, another for the confirmation. Given that the longest-running searches have been collecting data for just over a decade, they've only had time to hunt down planets that are at most three Earth-Sun distances from their central star. Jupiter, at just over five times the distance of Earth from the Sun, completes an orbit every 11.9 years, and would therefore take almost a quarter of a century to track around two circuits.

The upshot is that our present knowledge of extrasolar planets isn't anywhere near mature enough to provide support for the Rare Earth position.

If anything, it leans the other way. Consider: astronomers are confident that planetary systems in general are common. This belief stems from theoretical ideas about how planets form and from the observation that a high percentage of young stars are surrounded by dusty disks that contain the raw material for world-building. The one certainty we have is that only a small fraction—about one in twenty—of nearby Sun-like stars have big planets in small orbits. These planets are the *only* kind that, with our present detection methods, we can be pretty sure of not overlooking. The fact that they've shown up around only a tiny minority of stars is therefore positive news. As long as we remain effectively blind to planetary systems like our own, we have no data from which to conclude that they won't be widespread. Jupiters running amok and scattering smaller worlds like nine-pins may be a bigger factor than scientists suspected a decade ago (when they weren't contemplated at all), but it's much too early to claim that this is going to significantly curtail the prospects for advanced life.

As search techniques improve, their biasing effects will lessen, and we'll start to assemble a much clearer picture of what makes up a typical collection of planets. The trend away from finding only super-hefty, close-in gas giants is already starting to kick in as researchers report more extrasolar planets in the Saturn size range. Soon we'll be into the Neptune range. And Debra Fischer at the University of California, Berkeley, after looking closely at the data of the host stars of various known extrasolar planets, has found evidence of wobbles due to additional unseen companions in at least half the cases. On top of this, an entirely different approach to planet detection that relies on a phenomenon known as microlensing (of which more in Chapter 8) has provided tantalizing signs of Earth-mass planets around other stars.

As time goes on, not only more and more single worlds but increasing numbers of families of planets seem set to come to light. Only when we can compare many entire planetary systems, and see the role that smaller, rocky worlds like our own play in them, will we be able to make an informed judgment on the abundance or rarity of habitable Earth-sized planets in stable, circular orbits.

One further point needs mentioning, because it threatens to become a major issue over the coming months and years. In October 2000, George Gatewood, director of the University of Pittsburgh's Allegheny Observatory, David Black of the Lunar and Planetary Institute, and Inwoo Han of the Korea Astronomy Observatory presented tentative results suggesting that many known extrasolar planets may not be planetary at all. Instead they may be small cool stars known as brown dwarfs, or lightweight red dwarfs.

We mentioned above that the method normally used to hunt for extrasolar planets—measuring the extra back-and-forth movement of a star along our line of sight caused by an invisible orbiting companion—can only give an accurate *minimum* mass for the companion. Unless the orbit of the companion is seen edge-on, the object's actual mass will be higher, but this method doesn't allow the tilt of the orbit to be worked out. Another method of detecting invisible companions is to look for side-to-side wobbles in a star's movement—its motion at right-angles to our line of sight—by astrometry. This is more difficult because it involves measuring the actual changes in position of the star in the sky (as distinct from changes in the position of its spectral lines), which are exceptionally small. Nevertheless, thanks to the European spacecraft *Hipparcos*, high-precision astrometric data are now available for tens of thousands of stars, including those around which "planets" have been found. Gatewood, Black and Han examined the *Hipparcos* data together with the line-of-sight data used in 30 planet discoveries. They concluded that the orbits of most of these companion objects are nearly face-on to us; therefore, the masses of the objects are much larger than had been previously claimed.

According to these new results, only nine of the 30 stars studied have companions with masses of 10 to 15 Jupiters or less, that would put them in the planetary range. Above about 13 Jupiter masses, astronomers believe that an object would undergo some nuclear fusion in its core and glow dully in the infrared as a brown dwarf—the smallest, least massive kind of star. A second group of 11 stars appear to have companions with masses of 15 to 80 Jupiters, which would make them brown dwarfs. A third group of four stars yields companion masses above 80 Jupiters, placing them in the red dwarf range. A final group of six will require more astrometric observations before companion masses and orbital tilts can be confidently worked out.

These results are consistent with earlier studies that have noted that the orbital periods and eccentricities of so-called "extrasolar planets" are distributed in a way that is statistically indistinguishable from binary stars. It's an important issue because planets and stars differ in the way they form—planets from dusty disks circling around stars, stars from the collapse of interstellar clouds. At the time of writing, however, the issue is unresolved. So, for the sake of argument, we'll assume, as *Rare Earth* does, that all of the claimed planets are as advertised.

Jupiter's Protection

Another claim of Rare Earth advocates is that, in order to nurture advanced life, a planetary system must contain a large gas giant moving in a wide, circular orbit. Such a planet, the argument goes, serves as a bodyguard, deflecting asteroids and comets away from the inner regions and so preventing large numbers of these stray objects from crashing into any worlds on which complex life is destined to evolve. The Earth was extraordinarily lucky, it's said, to have Jupiter as a protector. This idea was originally put forward by George Wetherill, of the Carnegie Institution, in 1995.

First, it's worth mentioning that we already know of a few "classical" extrasolar Jupiters—gas giants occupying fairly large, not-too–elliptical orbits. For example, a planet going around the star 47 Ursae Majoris has a mass just over twice that of Jupiter, and another going around the star HD 10695 has a mass about six times that of Jupiter. Both orbit at distances similar to that of the asteroid belt from the Sun, in paths only slightly more elliptical than the orbit of Mars. In December 2000, the discovery was announced of a Jupiter-mass planet in a near-circular orbit, only slightly larger than that of the Earth, around Epsilon Reticulum. To find such relatively conventional gas giants at this early stage in the search suggests there may be a very large reservoir of them waiting to be found over the next couple of decades.

Second, this matter of Jupiter-like planets in Jupiter-like orbits is not divorced from the issue of Earth-like planets in Earth-like orbits. Yet Rare Earth advocates treat the two as statistically unrelated events, different rolls of the dice. This is a little like the movie character who, when told that Lou Gehrig died of Lou Gehrig's disease, asked, "Gee, what are the odds of *that*?" In fact, the odds are pretty good. If a planetary system hasn't been rearranged, early on, by big planets hurtling all over the place, you'd expect the original design still to be in place: small, terrestrial-type worlds close in and large, Jovian-type worlds further out. One implies the other. As we've seen, it's too early to say how common these relatively undisturbed planetary set-ups really are.

"Metallicity"

Much has been made, in the Rare Earth debate, of the issue of "metallicity." Astronomers, often a bit cavalier about chemistry, are prone to call any element that's heavier than helium a metal. So a star's "metallicity" is simply a measure of the proportion of elements it contains other than hydrogen and helium. This factor bears on planet formation, because heavy elements are needed for building the central parts of all planets (including gas giants) and the bulk of rocky worlds. The stuff available for making planets is what's left over from the formation of the central star; therefore, stars and their surrounding protoplanetary disks have very similar compositions.

The host stars of most of the extrasolar planets found to date have a fairly high metallicity: about four-fifths of them are at least as metal rich as the Sun. Rare Earth supporters suggest that this might seriously limit the number of planetary systems, since stars with high metallicity tend to be found only in the inner reaches of the galaxy. Guillermo Gonzalez has been especially vocal on this point. But once again, for a perfectly good explanation, we need look no further than the fact that the first extrasolar planets to be discovered have tended to be high-mass planets in unusual orbits. In general, we'd expect those Sun-like stars with the highest proportion of heavy elements to be the ones hosting massive planetary systems containing multiple large Jupiters. Where there's a glut of raw material available for assembling large, metallic and rocky planet cores in closely stacked orbits, the natural outcome will be a number of outsize gas giants moving close together. And then all hell will break lose. Interactions between nearby giants will scramble the planetary orbits, until you end up with a motley assortment of hot and eccentric Jupiters. Anomalously high metallicity and anomalously located gas giants will tend to go hand in hand, which is just what astronomers have found.

However, there seems to be no hard and fast rule. For example, the star 70 Virginis has almost the same metallicity as the Sun but boasts a planet with seven times the mass of Jupiter, while HD 114762, which is markedly subsolar in heavy element content, has a planet as massive as eleven Jupiters. Nor does there seem to be any consistent link between the type of orbit and metallicity. HD 37124 is orbited by an eccentric Jupiter, discovered in 2000 (after the publication of *Rare Earth*), and has the lowest metallicity of any star known to have a planet (below average, in fact, for stars of its age and type)—a sign perhaps of things to come. On the other hand, Gliese 86, which also has a much lower metallicity than the Sun, is accompanied by a hot Jupiter in a much smaller, nearly circular orbit. The picture is further complicated by the fact that some stars may have a heightened metallicity because they've swallowed one or more giant worlds. Only time, and a great deal more data, will help resolve this issue.

As a rough guide, around stars with a lower fraction of heavy elements, we'd expect the gas giants to be smaller (more like our own Jupiter and Saturn) and more widely spaced at birth so that they don't toss each other around and disrupt the initial arrangement. Because our current search methods can't pick up these conventional solar systems, the present data seem to suggest there's a cutoff of planets below a certain metallicity (though the cutoff value is rapidly dropping). But this is only because of the selection effect we've already talked about. It's an inevitable consequence of a search method still severely limited in sensitivity.

The trend of new planet discoveries toward lower-mass worlds and gas giants located farther from their central stars suggests that we've had most of the "bad" news concerning the likelihood of Earth-like worlds in Earth-like orbits. And, in fact, the news isn't so bad at all. We can now look forward to what most astronomers have long anticipated—the discovery of numerous terrestrial-type planets orbiting within their stars' habitable zones. If this turns out to be the case, then continuously active plate tectonics, which may (or may not, if Chris McKay is right) be crucial for the kind of advanced life we know, can be expected to be routine as well. Even within the solar system there's evidence that both Venus and Mars have had plate tectonics in the past. And if Venus had formed where Mars is, it might still have plate tectonics today. This issue of mobile crustal blocks isn't an additional factor to the existence of suitably placed planets.

The Galactic Habitable Zone

The last item on the Rare Earth laundry list is the importance to complex life of stars that move within the so-called "Galactic Habitable Zone." This, again, is a concept that Guillermo Gonzalez has heavily promoted. He and a handful of other astronomers have suggested that only stars which orbit at the right distance from the center of the Galaxy and at the right rate might have a chance of supporting higher forms of life. The "right distance" means close enough to have a sufficiently high metallicity for planet-building (though we've just seen the weakness of this constraint), but not so close to risk exposure to biologically unfriendly levels of radiation from the galactic core. The "right rate" means orbiting at roughly the same speed as the spiral wave motion that defines the location of the galaxy's bright arms. A star that plunges in and out spiral arms too often, the argument goes, will be dangerously subject to the effects of supernovae.

More than any other argument put forward in support of the Rare Earth cause, this smacks of someone having gone out of their way to find something unusual about the Sun and Earth. Where are the data from which we can draw conclusions about the effects of galactic radiation on a planet's biosphere? Where are the tables showing how supernovae, of different types and distances, affect evolutionary trajectories on worlds in their neighborhood? The Sun has almost certainly been exposed to supernovae at various ranges over the last few billion years. In fact the solar system is currently inside a large cavity of hot gas—the Local Bubble—that is thought to have been formed by one or more supernovae in the relatively recent past. Evidently, we've survived the experience. How do we know whether more supernovae, or radiation exposure in general, would have been good or bad for evolutionary progress? Mutations would certainly have occurred at a greater rate, generating more diversity and opportunities for natural selection.

The more you think about it, the more it seems there's something odd about the Rare Earth Hypothesis. At a time when many of our scientific indicators suggest, if anything, that life of every description, from the most primitive to the most complex, may be widespread, along comes this curious rebuttal. Of course, it's always good for the other side of an argument to be well-aired. Astrobiologists have a vested interest in playing up the evidence for extraterrestrial life, and it can only be healthy that doubters keep them honest. But the sudden rise of the Rare Earth position in the latter half of the 1990s, at precisely the time when astrobiology was taking off, and SETI projects were starting to sprout up everywhere, is really quite puzzling. At least, it's puzzling until you understand what is driving part of the Rare Earth campaign.

The idea that there's something very special about our planet has always been essential to those who maintain that Earth and Man are of divine origin. If, for instance, you opt for a strict Biblical version of reality, you're bound to accept that this planet and we ourselves are not natural in origin but the products of intelligent design. The prospect of other Earth-like planets, inhabited by other intelligent beings, casts doubt on the unique relationship between God and the human race.

This might seem irrelevant to astrobiology except as historical preamble. Yet surprisingly, at the dawn of the twenty-first century, religious influence is once again being brought to bear on the question of whether or not the Earth is somehow special. Without many people realizing it, debate in astrobiology is being actively manipulated by deeply held theological beliefs.

The revival of the Rare Earth debate—its latest reemergence—began in earnest just after the discovery of the first extrasolar planets around Sun-like stars in the mid-1990s. A few people started putting out what might be described as anti-SETI articles and papers. Alan Rubin, for example, produced a skeptical piece on extraterrestrial intelligence for the *Griffith Observer* in 1997. But when you look at the material written over the past few years, attacking the notion that planets and life of every description are probably common, one name crops up again and again.

Guillermo Gonzalez, whom we've met in connection with the size of the Moon, stellar metallicity and the Galactic Habitable Zone, is a young professor of astronomy at the University of Washington who, everywhere he looks, finds signs that the Earth is unique. Although he personally may not be well known, his writings and opinions have been widely disseminated in publications ranging from *The Wall Street Journal* to *Scientific American*. More to the point, he's played an important role in influencing Ward and Brownlee, whose book in turn is now exerting a powerful effect on public and, to some extent, professional opinion. One scientist who admits to having been swayed by some of *Rare Earth*'s arguments is the planet-hunter Geoff Marcy. "It's courageous," comments Marcy. "It's rare in literature and science that a stance goes so far against the grain." But Marcy, like the great majority of astronomers, is no Rare Earther: "For the first time in history, we've been able to prove that there are many planets orbiting other suns and there are probably many Earth-like analogs that have liquid water and atmospheres. There's no question that there are oceans and lakes out there. No question that the universe is teeming with life."

As Ward and Brownlee point out in their preface, "Guillermo Gonzalez changed many of our views about planets and habitable zones." "We often met," recalls Gonzalez, "during the last couple of years while the book was being put together to discuss astronomical constraints on advanced life." It would be going too far to say that Gonzalez pioneered many of the ideas in the Rare Earth Hypothesis. His main role as an innovator in the debate has been confined to the issues of stellar metallicity—his specialty—and the Galactic Habitable Zone. But more than anyone, he's been instrumental in tying together the various strands of the Rare Earth argument and energetically promoting the thesis across a broad front. While others have been caught up in new planet passion and Europan ecstasy, Gonzalez has sounded more than a note of caution. He has gone out of his way to seek and present evidence opposing the idea that we shall ever find other worlds with complex life and intelligence. It's a bold and lonely stance to take in the present climate of optimism in astrobiology, and Gonzalez is to be congratulated on stating his case in the face of overwhelming opposition. "It was not something I took lightly," he explains, "as there are several strong SETI supporters in our department."

I have a stack of articles on my desk that, if read rapidly in succession by an astrobiologist or SETI enthusiast, would probably have him or her on the phone to the Samaritans. It's all bad news, and it's either written by Gonzalez or quotes him as a primary source. Everything he comes across suggests to him that we're lucky to be here, that complex life on Earth is balanced on a knife edge. And Gonzalez is pretty sure he knows why.

I mentioned that I have a stack of articles by Gonzalez. In fact I have two stacks. The second makes even more surprising reading than the first, and probably wouldn't be on my desk at all if it hadn't been for an unusual concatenation of events. Briefly, this involved a wife who was looking for a church hall for her Girl Scout Troop and a pastor who, during the negotiations, discovered that this lady's husband was writing a book about life in the universe. Thus I came by a copy of *Connections*, a quarterly newsletter published by Reasons to Believe, Inc., of Pasadena, California, whose mission is "to communicate the uniquely factual basis for belief in the Bible." Not my usual literary fare, *Connections* contains articles that attempt to use (or usurp) scientific evidence to support the creationist cause. The first article in this particular issue (Volume 1, Number 4, 1999) was "Live Here or Nowhere," by Hugh Ross (the president of Reasons to Believe and a well-known creationist scientist) and a certain Guillermo Gonzalez. A brief check of the references and a few minutes on the Internet were enough to confirm that this was indeed Gonzalez the astronomer and chief Rare Earth campaigner. "Live Here or Nowhere" concludes with the sentence: "The fact that the sun's location is fine-tuned to permit the possibility of life—and even more precisely fine-tuned to keep the location fixed in that unique spot where life is possible—powerfully suggests divine design." One of the references in this article is to a paper by him (then in press) called "Is the Sun Anomalous?" in the scientific journal *Astronomy & Geophysics*. Oddly enough, this paper contains no mention of divine intervention.

A little more research reveals that Gonzalez has been living something of a double life, producing standard scientific output on the one hand and, more or less simultaneously, penning other articles on similar topics in which the same conclusions are presented solely for the purpose of supporting the design argument. As a regular contributor to Reasons to Believe pamphlets, he is no mincer of words. In a 1997 piece, he writes:

> I see no way for life, unless governed by a super-intelligent Creator, to pre-
> dict and respond perfectly to ongoing changes in the other balanced fea-
> tures. Life is so information rich and its environment so narrowly defined
> as to defy strictly natural explanation. The personal involvement of a
> supernatural Creator seems scientifically reasonable to me.

A 1998 article, "Design Update: How Wide is the Life Zone?" has him firing this opening salvo:

> Two decades ago, researcher Michael Hart provided an important piece of
> evidence in the argument for divine design. Using long-term climate simu-
> lation models, he showed that the region around the sun in which a rocky
> planet . . . can continuously support life . . . is a very narrow band. . . . Two
> years ago, however, Hart's conclusion was challenged by James Kasting . . . .
> [who] estimated that the CHZ [the continuously habitable zone] around our
> sun is much larger . . . Those with a Christian world view can still see the
> miracle of our location, but those with a nontheistic perspective seemed to
> gain some room for happenstance. [A] further look at Kasting's estimate
> shows that they gained nothing, except perhaps the opposite of what they
> expected, a further accumulation of evidence for design.

In conclusion, he writes, "Scientists are left to wonder how Earth came to exist and persist for so long in the zone where life is possible. The impression of design could hardly be more distinct."

Of course, it isn't unusual for professional scientists to hold strong religious beliefs. Most of the time, it isn't relevant or, frankly, anyone else's business. But in the case of Gonzalez it matters because his underlying conviction has led him to play a very significant role in raising the prominence of the Rare Earth debate. "Gonzalez has been a big influence," commented Ward.

In such a situation, where researchers are working side by side, sharing ideas and co-authoring papers, one would assume that their wider beliefs become common knowledge among the group. Surely, Ward and Brownlee were aware of their colleague's deep theological convictions. But an inquiry to Ward on this issue in August 2000, eight months after the publication of *Rare Earth*, triggered a wholly unexpected and rapid exchange of e-mail communications:

DARLING TO WARD: I must ask you one other question because it arises in the book, although it's on a touchy issue. Again, it regards your collaboration with Gonzalez and the influence he's had on the Rare Earth debate. As I'm sure you're aware, G. writes extensively as a Christian apologist and uses the same arguments he presents in scientific form—the Galactic Habitable Zone, etc— to support his case for divine design. My question is, do you think there is any chance that the arguments being put forward by you and others, purely on a scientific basis, have been influenced and possibly biased by G's "hidden" agenda?

W TO D: That is news to me—I have never seen or even heard that Gonzalez does this—we [Gonzalez, Ward and Brownlee] are writing a huge paper for *Icarus* on metallicity and there has never been a whisper of intelligent design, in the two years I have known him. Are you sure you have the right Gonzalez?

D TO W: There is no doubt.

W TO GONZALEZ (having forwarded the above messages): I think I need an explanation.

G TO W: Regarding his statement about my "secret agenda" as a design advocate, it is not such a secret, as my writing on the design issue is rather public and widely distributed. I recently received a grant from the John Templeton Foundation to study habitability from a design perspective—several people in the department know about it. I have not been more open about my pro-design views here at the UW because of the open hostility to such views among many faculty. But, I certainly will not apologize for admitting that my theistic theological views motivate my science and vice-versa.

So here is a curious situation of a scientist actively seeking evidence that extraterrestrial life is rare to shore up a belief in divine design. And doing it, moreover, without the knowledge of many of his peers, who are nevertheless being strongly influenced by work that is intrinsically biased. Yet it isn't without precedent; much science, both good and bad, has arisen from motivations outside science. Copernicus, Einstein, and the modern cosmologists who postulate a "flat" universe have all been influenced in important ways by aesthetics; so too have particle physicists (as Brian Greene makes clear in his aptly-titled book *The Elegant Universe*). Much of nineteenth-century anthropology was explicitly racist. And Louis Agassiz's great expeditions to South America in the 1860s were undertaken in an overt attempt to shore up creationism against Darwinism.

Agassiz, who had arrived in the United States in 1846 and became professor of zoology at Harvard, was the last American biologist of any note to reject evolutionary theory following the publication of Charles Darwin's *The Origin of Species*. For him, the divine plan of God revealed itself throughout nature, and he was determined to find evidence to prove his case. In 1865, he set out for Brazil to collect specimens of fish from isolated pools in the Amazon basin. Since all the pools were essentially identical in their physical environment, he reasoned, Darwinian evolution ought to produce exactly the same adaptations and therefore exactly the same species of fish in each pond. The fact that the fish differed quite widely he took to indicate a Creator who delighted in diversity. This conclusion reflected a profound misunderstanding of Darwin's theory.

Yet Agassiz's work in Brazil was not wasted. Prior to his expedition, scientists knew of only about 100 species of fish in the whole of South America. The specimens he brought back to Harvard increased that number more than twentyfold, and their glorious diversity came to be understood as an important confirmation of evolution. They can still be seen in the Peabody Museum of Comparative Zoology, where they helped build the collection of what remains today one of the world's great institutions for the study of evolution—an institution, moreover, whose founder was Louis Agassiz.

There's nothing intrinsically wrong with starting from an assumption of intelligent design and setting out to prove it scientifically. In the end, a scientist's personal beliefs are irrelevant because science is a communal, peer-reviewed venture in which belief is always subordinate to data. By drawing so much attention, the Rare Earth Hypothesis has laid itself, and the case for intelligent design, open to scientific scrutiny. As John Stuart Mill pointed out, nothing will expose the weaknesses of a bad idea (or, one might add, the strengths of a good one) faster than its wide dissemination. With the issue in plain view, it now comes down to observation.

Only time, and much more research, will tell who is right. Perhaps those who seek evidence of the hand of God will find what they're looking for. And if, after long seeking, we discover that the Earth and complex life and intelligence *are* rare, or even unique, then we'll have to consider whether this might be due to good luck or good management. For the moment, there's really no case to answer. Beyond the selection effects and hidden agendas that offer the last bastion of hope to those who still cling to the belief that somehow we are privileged, the Copernican revolution that began more than four centuries ago is quietly running its course. Nowhere is that more evident than in the way life has evolved on our own world.

7
Theme and Variation

Life not only appeared on Earth, it evolved into many millions of different species, one of which can even ask questions about itself and the universe. How did it work, this long unfolding and synthesis from microbe to microbiologist? What pivotal influences guided evolution here? Would those same influences shape the nature and appearance of life elsewhere?

The physicist Guiseppe Cocconi, an early pioneer of SETI, remarked, "This probing of evolution is really a fantastic thing because we know only of one way evolution worked, on Earth. It is one history, our history. But there are probably millions of other roads . . ." The problem of sample size—the lack of a second known living planet—is the bane of contemporary astrobiologists. But whereas life on Earth represents a single entry in an otherwise empty logbook, evolution offers us many different creatures to study. All may be of common stock, yet they give us a healthy basis for seeing how the forces of evolution play out over long periods and in different ecological settings. The dazzling yet bounded variety around us contains clues to the kind of life we can expect on other worlds.

☼

Evolutionary theory has itself evolved and continues to do so, as competing ideas battle for scientific supremacy. Its modern story begins in late eighteenth century France with Jean-Baptiste Lamarck, keeper of the royal garden and later professor of invertebrate zoology at the Museum of Natural History in Paris. In his *Philosophie zoologique,* published in 1809, Lamarck put forward the first theory in which organisms were seen to change and evolve by a gradual process of adaptation to their environment. According to his scheme of "inheritance of acquired characteristics," small physical changes that creatures picked up during their lifetimes were handed on immediately to their offspring. Giraffes that stretched a bit further for their food, for example, might pass their slightly longer necks on to the next generation.

Lamarckism, which remained popular in France for most of the nineteenth century, encouraged the first speculations about deeply alien life. Although writers had long fantasized about the sorts of creatures that might inhabit the Moon and other worlds, these early extraterrestrials were little more than extravagantly dressed humans and chimerical animals cobbled together from terrestrial body-parts. The idea that creatures adapt subtly to suit the particular environment in which they live encouraged some of Lamarck's contemporaries to think more imaginatively about what kinds of life might emerge on worlds very different from Earth.

One of these imaginers was the French astronomer Camille Flammarion, whose popular writings did more than anyone else's to stimulate public interest in the heavens. Flammarion believed passionately that life was common beyond Earth, and wrote at length in his best-selling work, *Astronomie populaire*, first published in 1880, about the biological prospects of Mars and even the Moon. It was in his fiction, however, that he gave himself free rein. *Real and Imaginary Worlds* (1864) and *Lumen* (1887) conjure up a range of exotic species, including sentient plants that combine the processes of digestion and respiration.

Even more *avant-garde* were the astrobiological ruminations of the Belgian writer Joseph-Henri Boëx, writing under the pseudonym J. H. Rosny the Elder. In his first novel, *Les xipéhuz* (The Shapes), published in 1887, he describes the arrival on Earth in prehistoric times of strange translucent beings that threaten the survival of the human race. Other compellingly bizarre life-forms appear in his *Un autre monde* (Another World, 1910) and *La mort de la terre* (The Death of the Earth, 1910). It's been said that the allure of Rosny's speculations was a factor in Jules Verne's decline in popularity in the 1890s, even though Verne is incomparably better known today.

While Lamarckism remained strong throughout the nineteenth century in its country of origin, elsewhere it gave way to a new, more insightful vision of how life evolves. In England, two young, somewhat eccentric Victorian country gentlemen with time on their hands set about documenting the minutiae of the species around them. Alfred Russell Wallace and Charles Darwin both began their studies in the quiet English countryside, both sailed to the tropics in the mid-1800s to further their research, both returned with data that would ultimately convince them that species diverge, over time, from common ancestral stock as they adapt to local environments. Both eventually arrived independently at the same conclusion, having been inspired by the Reverend Thomas Malthus' book *Essay on the Principle of Population*, first published anonymously in 1798. Malthus pointed out that populations grow faster than the food supply available to them. Whenever resources are limited, certain traits prove more conducive to survival than other traits. Nature acts a selective force, killing off the weak and forming new species from the survivors, who are by definition better fitted to their environment. This, of course, is the basis of Darwin's theory of evolution by natural selection.

Why isn't it the "Wallace-Darwin" theory? Both men went public with their conclusions simultaneously, by common consent, at the same meeting of the Linnaean Society of London in 1858. Why is Darwin remembered and not Wallace? For the simple reason that Darwin wrote a book about the new theory, *The Origin of Species*, which created a sensation as soon as it appeared in 1859.

Like Lamarckism, Darwinism had a particularly potent influence in its nation of birth, the more so since it was championed tirelessly by the distinguished biologist Thomas Huxley and the philosopher Herbert Spencer. Spencer coined the phrase "survival of the fittest" and stoutly defended the view that evolution wasn't peculiar to Earth, but was a universal imperative. In his opinion, the cosmos had a natural tendency to generate order from chaos and drive the development of higher forms of life from lower ones. In a sense, he foreshadowed the commonly held view that life emerges almost inevitably, given the right starting conditions.

Huxley was more of a pragmatist, and a brilliant and persuasive orator. His 1868 Edinburgh lecture "On the Physical Basis of Life" was the spark for the scientific debate about the chemical origin of life. In it, he said:

> I suppose that to many, the idea that there is . . . a physical basis, or matter, of life may be novel. . . . If the properties of water may be properly said to result from the nature and disposition of its component molecules, I can find no intelligible ground for refusing to say that the properties of protoplasm result from the nature and disposition of its molecules.

Huxley also exerted a decisive effect on the mind of a young writer who attended his lectures at the Normal School of Science (later renamed the Royal College of Science) in London. Like many of his contemporaries, H. G. Wells came to appreciate that evolution was a central principle of life.

This thinking became a focal point of much of his fiction and science journalism. In essay after essay, especially in his first decade of professional writing from 1887 to 1896, he attacked the traditional anthropocentric view that man was somehow privileged and predestined. What was *Homo sapiens* but another episode in the panoramic sweep of history? From that leaping-off point, he went on to contemplate the precariousness of man's tenure on Earth. In the early piece "Zoological Regression," he writes:

> There is . . . no guarantee in scientific knowledge of man's impermanence or permanent ascendancy. . . . [I]t may be that . . . Nature is, in unsuspected obscurity, equipping some now humble creature . . . to rise in the fullness of time and sweep *Homo* away. . . . The Coming Beast must certainly be reckoned in any anticipatory calculations regarding the Coming Man.

In *The Time Machine,* the Coming Beast turns out to be mankind itself— or rather, two degenerate strains descended from the British aristocracy and working class. Then, in April 1896, in a *Saturday Review* article called "Intelligence on Mars," Wells offered another possibility. What would sentient life be like on the fourth planet? With Lowell and his canals at their popular zenith, it was an issue of the moment, at least in the layperson's mind. But Wells poured scorn on suggestions that the inhabitants might resemble ourselves:

> No phase of anthropomorphism is more naïve than the supposition of men on Mars. The place of such a conception in the world of thought is with the anthropomorphic cosmogonies and religions invented by the childish conceit of primitive man.

The Martians, he concluded, "would be different from the creatures of earth, in form and function, in structure and in habit, different beyond the most bizarre imaginings of nightmare." But not, as it turned out, beyond the imagining of Wells. A year later, in *The War of the Worlds,* he unleashed one of his darkest evolutionary visions, a tale that was to exert such an irresistible grip on the public conception of alien life that its effects are still felt today. The fearful thought of a malign intellect "vast and cool and unsympathetic," when broadcast as a radio adaptation in 1938, led to mass panic. (It was an unrelated Welles (Orson), who had the unfortunate idea of putting the adaptation in a news-documentary format.) Later, the same fear helped fuel the flying saucer myth and ensured endless fascination with horror-alien depictions, from *It Came From Outer Space* to the *Alien* tetralogy.

Wells was a "divergionist," meaning that he believed the overriding thrust of evolution is to generate new, unique forms—to explore novel options and possibilities, never venturing down the same taxonomic byway twice. He'd learned under Huxley that if you change the circumstances, you change the way life adapts. Wells simply extrapolated that reasoning out into the universe. Since no two worlds can have exactly the same environment, novelty and endless variety must be the name of the evolutionary game across space and time.

The unrepeatable nature of evolution became a common idea among biologists, and it was further encouraged by what was perceived to be an almost unbelievably fortuitous chain of events that led to the origin of life on Earth. As soon as scientists in the 1920s began looking at the details of the chemistry that might have given rise to the first organisms, they started to lean toward a couple of conclusions. One was that evolution would never happen the same way twice—something that Alfred Wallace had first suggested. Even if you started with the same ingredients in the same proportions under exactly the same conditions, you'd get a completely different biological outcome—assuming you got a biological outcome at all. And that was the other point. The chemical origin of life seemed to depend on such an improbable sequence of events, similar to throwing a die over and over and getting a six every time, that biologists were inclined to think that life elsewhere must be a very rare occurrence.

During the first half of the twentieth century, biologists seldom got involved with the question of extraterrestrial life. There were too few data to work with and, in any case, they had enough to do on Earth. Most of the speculation about life in the universe came from astronomers, who were generally positive about the idea simply because they thought there were probably so many planets around. With billions of potential homes, surely life couldn't be that scarce.

One of the few life scientists in the early twentieth century who commented openly on the prospects for finding alien organisms was William Diller Matthew. A paleontologist at the American Museum of Natural History in New York from 1895 to 1926, Matthew later became head of the Paleontology Department at the University of California, Berkeley. He found it "noticeable that, as usual, the astronomers take the affirmative and the biologists the negative side of the argument." Life had arisen on Earth, he said, as a result of "some immensely complex concatenation of circumstances so rare that even on earth it has occurred probably but once during the eons of geological time." As for extraterrestrial life, if it existed at all, "it probably—

almost surely—would be so remote in its fundamental character and its external manifestations from our own, that we could not interpret or comprehend the external indications of its existence, or even probably observe or recognize them."

In their hugely popular 1931 encyclopedia, *The Science of Life*, H. G. Wells, his son G. P. Wells, and the biologist Julian Huxley (grandson of Thomas) coined a name for such completely alien biota: "Beta Life." It would be

> . . . an analogous thing and not the same thing. It may not be individualized; it may not consist of reproductive individuals. It may simply be mobile and metabolic. It is stretching a point to bring these two processes under one identical expression.

Twenty years later, Harold Blum, professor of biology at Princeton University, speculated along the same lines in his *Time's Arrow and Evolution*. If life exists at all elsewhere in the universe, he wrote, "it probably has taken quite a different form. And so life as we know it may be a very unique thing after all, perhaps a species of some inclusive genus, but nevertheless a quite distinct species."

Today, divergionism, as a school of thought in evolutionary biology, goes under the name of "contingency." Perhaps its best-known contemporary standard-bearer is the Harvard paleontologist Stephen Jay Gould. In his 1989 book *Wonderful Life*, he argues against the repeatability of species, and of humanoids in particular, in a style and manner that are purely Wellsian. He points to the

> staggeringly improbable series of events, sensible enough in retrospect and subject to rigorous explanation, but utterly unpredictable and quite unrepeatable. . . . Wind back the tape of life to the early days of the Burgess shale; let it play again from an identical starting point, and the chance becomes vanishingly small that anything like human intelligence would grace the replay.

This Burgess Shale of which Gould writes contains a collection of some of the best preserved and oldest invertebrate fossils ever brought into the light of day. It reveals in extraordinary detail the extent of natural biological trials that went on at the dawn of the Cambrian era, during the rise of complex animal life, some 530 million years ago. The Cambrian Explosion, in fact, is sometimes referred to as the Cambrian *Experiment*. In the Burgess shale we find exposed, as nowhere else, the remarkable breadth of possibilities that were explored at this time. Turning over the thin pages—the closely-spaced lamina—of this clay rock, you see the imprints of life-forms so utterly unfamiliar that you might imagine yourself a fossil-hunter on the other side of the galaxy.

Modern arthropods—the insects, spiders, and crabs—are constructed around just three different body-plans: the Burgess fossils show that arthropods alive in the Cambrian were based upon at least *twenty-four* distinct body arrangements. Another 10 to 15 animals are so bizarre that they've defied classification at all. One creature, for instance, sported five eyes and a nozzle as well as a mouth. Another had a mouth that seems to have worked like a nutcracker. Yet another, equipped with seven pairs of struts, has been aptly named *Hallucigenia*.

These fantastic animals left no descendants. Yet, Gould insists, there's no obvious reason why not. There's nothing to suggest that they were inferior to the animals whose lineages continued. Their extinction was no fault of their own, he concludes, but purely a matter of historical chance—of contingency. "Replay the tape of life," as he puts it, and the odds are that the insignificant little chordate worm that represented the earliest rendering of our own body plan would have fallen by the wayside. Something else would have taken its place. And, as a result, there would have been no humans.

This becomes a compelling viewpoint when you consider the mesmerizing diversity of life on Earth today. Think how incredibly different a hummingbird is from an oak tree, or a whale from an amoeba, and imagine all the creatures, different from anything we know, and the creatures that never descended from them, that might have been alive now but for some chance misfortune. Think about the long, complex chain of happenstance that must lie behind every single species with which we share this planet. Then consider how easy it would have been for the history of each life-form to have taken a different course at any point, and so given rise to an entirely new type of creature. When you reflect on this, divergionism or contingency, it seems clear, must be an intrinsic aspect of the way life unfolds.

Yet surely not the only aspect. Another school of thought in evolutionary biology centers not on divergionism, but on *con*vergionism. And if we've appointed Stephen Gould to be the spokesperson for the former, we can do no better than choose Simon Conway Morris, professor of evolutionary paleobiology at the University of Cambridge, as chief protagonist for the latter. It was Conway Morris who, along with his Cambridge colleagues Henry

Whittington and Derek Briggs, spent twenty years reconstructing the Burgess Shale fauna. In his 1998 book *The Crucible of Creation,* Conway Morris takes Gould to task for his emphasis on the role of contingency:

> [D]espite the length of the argument in *Wonderful Life,* its main strand concerning historical contingency . . . can be briefly explained. . . . Any historical process . . . must be riddled with contingent events. Their effect Gould maintains, is to render almost any prediction of the future course of history a futile . . . exercise . . . Gould argues passionately that were we "to replay the tape of life" from the time of the Cambrian explosion, we would end up with an utterly different world. Among its features would be an almost certain absence of humans or anything remotely like us. [But] this whole argument . . . is based on a basic confusion concerning the destiny of a given lineage . . . versus the likelihood that a particular biological property or feature will sooner or later manifest itself as part of the evolutionary process.

Conway Morris points out that biologists have long recognized *convergence* as a ubiquitous property of life. In his great work *On Growth and form* (1917), the British naturalist D'Arcy Thompson argued persuasively that unrelated animals and plants would develop the same characteristics—the same solutions—to common environmental challenges. He wrote, "[I]n general, no organic forms exist, save as are in conformity with physical and mathematical laws."

He believed that a new unifying vision was needed in biology:

> The search for differences or fundamental contrasts between the phenomena of organic or inorganic, of animate or inanimate things, has occupied many men's minds, while the search for commonality of principle or essential similarities, has been pursued by few; the contrasts are apt to loom too large, great though they may be.

Thompson observed that there are only a limited number of generic shapes that nature keeps using, and that these shapes are greatly modified by slight variations in their environment during growth and development. He drew two different species of crab, and superimposed a grid over each. By mathematical transformation of the coordinates he showed how easily a wide variety of crab shapes can be produced from just one archetypal crab plan.

Thompson pioneered the idea, familiar to all biologists today, that life keeps hitting, time and again, on the same solutions to the same basic problems of survival. Eyes, for instance, have been discovered, or reinvented, perhaps as many as 50 times in creatures as distantly related as mammals, cephalopods (the octopus eye is eerily human-looking but of vastly different ancestry) and insects. It's true that the morphology isn't identical. In vertebrates, the optic nerve comes into the eyeball at a certain point, and the nerve fibers spread out across the surface of the retina. Each individual nerve fiber reaches its assigned point, burrows down into the retina through several layers of epithelial cells, and ends with the light receptor itself pointing *away* from the lens. In the cephalopod eye, the optic nerve spreads out under the retina, and each nerve burrows up through the retina and ends with the light sensor on the surface of the retina, pointing *toward* the lens—a more efficient arrangement because it means there's no attenuation of light before it reaches the active components. Still, in both the vertebrate and cephalopod eyes, images are focused through a protein lens into cells filled with rhodopsin, a light-sensitive pigment that both types of animal have borrowed from plant cells. So the general eye structure and function is similar in two quite distantly related groups, the Chordate and the Mollusca. The chemistry and physics of vision are common, and in both cases determine the structural elements.

Wings or their equivalent are found in birds, bats, ptersosaurs, flying insects, rays and teleost fish. And speaking of animals that swim, Conway Morris notes that

> . . . there are only a few fundamental methods of propulsion. It hardly matters if we . . . illustrate . . . by reference to water beetles, pelagic snails, squid, fish, newts, ichthyosaurs, snakes, lizards, turtles, dugongs or whales . . . the style in which the animal moves through the water will fall into one of only a few basic categories.

Given that the world is subject to the laws of physics, natural selection is bound to keep discovering the same best strategies for living things to take advantage of these laws. The simple fact is that rhythmically beating wings and fins, wriggling bodies, and streamlined shapes are ideal for moving with the minimum of effort through a fluid.

It isn't just at the level of gross anatomy that biological convergence crops up. Examples of it have been found deep within the biochemical substrata of life. So-called "structural convergence" seems to be the only way of explaining the baffling similarity between certain antibody proteins in camels and the decidedly non–camel-like nurse shark. In most of the animal kingdom, the antibodies of the immune system consist of two chains, known as heavy and light, each chain having three loops. This arrangement, it turns out, is ideal for letting an antibody dock with and destroy an invading particle, such as a virus. But for some reason yet to be fathomed, some of the antibodies in camels and nurse sharks have lost their light chains. The surprising thing is that the response of both the camel and the nurse shark to this deprivation has been exactly the same. Both animals, in the course of their evolution, have increased the size of one of the loops—in fact, the very *same* loop—in their heavy chains. The modified antibodies in the camel and the shark act alike and look alike from a structural standpoint. But they're *genetically* distinct. Different nucleic acid sequences code for the antibody proteins in each animal.

In another case, biologists have come across a seemingly genuine instance of "sequence convergence" involving two groups of unrelated fish: the northern cod of the Arctic and the so-called notothenioids of the Antarctic. Both groups employ a kind of natural antifreeze to counter the effects of the icy waters in which they swim. Cheng Chi-hing and her co-workers at the University of Illinois have found that identical proteins make up this antifreeze in both groups. It's a straightforward structure based on the repetition of just three amino acids—analine, threonine and proline—over and over again. This informational simplicity, says Cheng, explains how it was possible for nature to stumble across the same sequence at completely different times and places: first in the southern polar region some seven to fifteen million years ago, then at the other end of the Earth several million years later.

So, all things considered, which is the more important influence in evolution, contingency or convergence? Does Gould win the debate, or does Conway Morris? No evolutionary biologist thinks this is an easy choice. The two effects of contingency and constraint, of fortune and physics, clearly work together, and it's a question of where the balance point lies between these opposing forces. Gould acknowledges the importance of convergence. Conway Morris is as aware as anyone, having been among the first to appreciate the spectacular diversity in the Burgess Shale, that life goes through intense periods of trial and error. Perhaps, in the end, that's the difference between the two. Gould might call it simply "trial" and leave it at that: there are trials of life, and which species live to fight another day is largely a matter of luck. Conway Morris, on the other hand, would include "error" too, because some creatures—most creatures, in fact—that arise during bursts of experimentation are not, in the long run, cut out to survive. For Gould, "error" doesn't exist; there are no maladapted organisms, only suboptimal environments.

Convergionism, while acknowledging that chance and circumstance play a part in shaping how life evolves in detail, lays greater stress on the fact that natural selection is subject to universal laws. Therefore the same motifs—the same anatomical and other adaptations—will recur in subtly different forms over and over again, as evolution finds the same solutions to the problems the environment throws its way. Gould is right in saying that contingency makes it impossible to predict which *species* will be favored over evolutionary time. In a rerun of history, there'd never be a recurrence of the same species we see around us today, including *Homo sapiens.* But Conway Morris replies that this isn't what really matters. Evolution becomes predictable, not at the species level, but rather in the timeless designs that natural selection keeps arriving at again and again. As Conway Morris puts it, "There may be little new under the Sun. This planet shows that despite the richness of life, the dance repeatedly returns to common themes."

This is the key we need for speculating about the nature of extraterrestrial life. Whatever forms aliens may take, convergence—ultimately, physical laws—will ensure that there's less deeply distinct variety and novelty than unbridled imagination would allow. There may be little new under other stars as well.

✲

Most scientists today, unlike their predecessors, no longer believe the origin of life was unlikely. Many steps may be involved, but given the extraordinary speed with which they happened on Earth, they seem almost inevitable if conditions are anywhere near favorable. The main point of debate today—and this applies to every phase in the emergence and evolution of life—is how *different* life might be.

How different might it be at the grassroots level of chemistry and biochemistry? Of course, we can't just dismiss weird and wonderful alternatives out of hand. Life based on silicon, boron, ammonia, methanol or any analogous substance you care to mention, has to remain on astrobiology's "improbable but not crazy" list. So too does life of those other intriguing kinds that fertile imaginations have dreamed up over the years. As long as there's uncertainty about how life began and evolved on our own planet, and while the star-fields of the galaxy remain unexplored, we can't delimit the outer bounds of what might be waiting for us on other worlds. All that astrobiologists can do at this stage, leaving aside the fantastic and the imponderable, is try to discern what seems most likely to be true about the majority of life throughout the cosmos.

The consensus view is that we'll encounter the pairing of carbon and water routinely at the chemical foundation of living things. Blame that partly on carbon-and-water chauvinism if you will, but the fact remains that these substances have no close rivals in their sheer number of biologically useful properties. The question then becomes, how far up the ladder of complexity do we have to go before we're likely to run into significant differences between life here and elsewhere?

There's a growing suspicion among scientists that many familiar carbon molecules are also essential parts of the universal constructor kit of life. Take amino acids, for instance—the repeating units of which proteins are composed. As we saw in Chapter 3, these can be manufactured extraterrestrially, in star-forming dust clouds, and so are available to form the basis of life at a wide variety of sites. But it's not only availability that makes them essential— we don't know of anything that could replace them in living things. On the other hand, more than 100 different amino acids occur in nature, and only twenty take part in making proteins in terrestrial organisms. Does that mean there's something special about these select few?

An experiment done by Andrew Ellington and his colleagues at the University of Texas in 1998 has shed light on this issue. These researchers grew a strain of the bacterium *Escherichia coli* that couldn't manufacture the amino acid tryptophan and so had to be given it as a nutrient. They also slipped into its feeding bowl the related synthetic amino acid fluorotryptophan, which is normally toxic to earthly creatures. With 100 percent artificial substitute, sure enough, the bacteria died within three cell divisions. But with 95 percent fluorotryptophan and 5 percent normal tryptophan, the *E. coli* survived and slowly grew. After many generations, they started to divide faster, as if mutations had sprung up that weren't as susceptible to the synthetic chemical's noxious effects. Eventually, the bacteria were able to cope with an exclusive diet of artificial substitute—multiplying very slowly but nevertheless surviving on their "alien" food. If terrestrial microbes can adapt so quickly to using a foreign amino acid, it suggests that unearthly combinations of amino acids might be commonplace among organisms with a different evolutionary history.

The idea that, with minor variations, life always ends up employing the same kinds of chemicals might seem to suggest a failure of imagination. Imagination, however, isn't the issue. It's relatively easy to invent fictional aliens whose blood resembles cleaning fluid or whose cells are filled with alcohol. The problems start when you have to fill in the biochemical details of how it would all work. Then it becomes clear there may be only so many ways of assembling elements into the kinds of materials that might be biologically tenable.

At the low end of the scale, where the first simple steps toward life are being taken, a lot of different chemical pathways and combinations of substances may come into play. Like a kid with a chemistry set, nature throws together every material at its disposal in every possible way. Most of these mad experiments fail, but a tiny fraction of them, by pure chance, result in something that resembles a primitive metabolism, a bit of a biochemical factory, or a half-baked means of self-reproduction.

At this stage, things aren't critical. It doesn't matter, for instance, if a certain proto-metabolic chain of reactions is hopelessly inefficient, because nothing alive yet depends on it and there are no predators to take advantage of any weakness. There's no reason why many such chains shouldn't be tried out all at once at different places on a young planet. Some will never advance far along the road to life because of inherent limitations. But those that do make progress and become more widespread will eventually start competing with one another for resources. At that point, Darwinian evolution kicks in.

Only the most successful protobiological systems, dancing at the edge of life, will win through to the next round. To this point, the dominant trend has been divergence—a riot of unrestrained experimentation. Now convergence starts to be a factor, guiding nature toward those chemicals, reactions and pathways that, cosmos-wide, do the job most efficiently and effectively. By the time the first organisms appear, scientists suspect, the choice of what works—what is biologically viable—may be very limited indeed.

Think about how a living thing gets its energy. No other chemicals are even remotely as well suited to storing and making energy available for life as carbohydrates—sugars and starches. As with amino acids, it's hard to see what could realistically take their place. This is a matter of circumstance as well as structure. The recent discovery of a simple sugar molecule, glycoaldehyde, in a large star-forming cloud near the center of our galaxy suggests that, like amino acids, sugars may be universally available in young planetary systems.

Still, while they may be irreplaceable as an onboard fuel supply for life, an organism can't simply burn, or oxidize, sugar directly whenever it needs a slug of energy. This would release *too much* energy all at once. Some way is needed of dividing up the energy in sugar into smaller amounts so that it can be tapped in a more measured way. The solution is for an organism to oxidize sugar and then transfer the energy to the chemical bonds of a high-energy compound. Once again, this imposes tight restrictions on life's biochemical options. Only three classes of sulfur compound and one of phosphorus are known to act as energy carriers in a biological setting. In terrestrial cells, the main power-pack for running the reactions of life is adenosine triphosphate, or ATP.

Energy must be transferred continuously from sugar to the energy carrier, so that plenty of charged-up molecules remain on hand to satisfy an organism's energy needs. On Earth, this involves the citric acid cycle (described in Chapter 2). Some biologists believe this may be another universal aspect of life's chemistry—perhaps among its most fundamental. "When we get to some other planet and find life," says Harold Morowitz of George Mason University, near Washington, D.C., "I have no idea what that life will look like, but it will have the citric acid cycle." To back up this claim, in 1998 Morowitz and his colleagues used a computer to sort through a database of 3.5 million organic molecules, looking for ways to convert carbon dioxide and water to citrate—a critical compound in metabolism—via molecules that incorporate only carbon, hydrogen, and oxygen. Applying six simple rules to identify suitable intermediate molecules, they whittled down the huge database to a short list of just 153 molecules, which included the eleven compounds involved in the actual citric acid cycle. Evidently, these eleven are far from a random set. What's more, the researchers used only a few obvious selection rules; additional rules might further reduce the number of realistic molecules for making citrate.

Living things may also routinely employ chlorophyll—the green, magnesium-containing substance that traps light energy and is at the heart of photosynthesis in plants on Earth. Being a biggish molecule, with well over 100 atoms, chlorophyll is not the sort of substance one imagines popping up all over the place. But scientists think there's a good chance that it's yet another of life's standbys. The reason, oddly enough, is that chlorophyll isn't ideally suited to capturing sunlight. The Sun radiates most strongly in the yellow region of the spectrum, whereas chlorophyll absorbs most strongly in the red and blue regions. You might suppose that the premier photosynthetic molecule on Earth would be a specialized yellow-light absorber. The fact that it isn't suggests that no better molecule exists for this purpose than chlorophyll. As the biochemist George Wald pointed out:

> Chlorophyll . . . possesses a triple combination of capacities: a high receptivity to light, an inertness of structure permitting it to store the energy

and relay it to other molecules, and a reactive site equipping it to transfer hydrogen in the critical reaction that ultimately binds hydrogen to carbon in the reduction of carbon dioxide. I would suppose that these properties singled out the chlorophyll [molecules] for use by organisms in photosynthesis in spite of their disadvantageous absorption spectrum.

If that's the case, chlorophyll is likely to evolve again and again wherever photosynthesis comes about.

❈

If many of the root substances of life prove to be common across the universe, as astrobiologists increasingly suspect, there could be little novelty among complex molecules either, since the same polymers will automatically assemble from the same monomers. Once amino acids, for example, are assumed to be a cosmic common denominator of life, the ubiquity of proteins must surely follow. Alien proteins will differ in detail from their terrestrial counterparts, but they'll be proteins nonetheless.

To direct the assembly of proteins, and of entire organisms, some kind of blueprint molecule is essential. Since Darwinian evolution is impossible without self-reproduction, there has to be a genetic substance of some kind.

Why, though, should we assume that evolution is universal? It's easy to imagine living things that don't evolve. One possible variety would be immortal beings. But without evolution in the first place, there's a problem understanding how an eternal life-form could come into existence. We can also imagine creatures that don't evolve because they always, unfailingly, make perfect copies of themselves—an endless lineage of identical clones. Such a non-evolver, in fact, seems more realistic that an organism that lives forever. It's even conceivable (though very unlikely) that primitive perfect copiers do sometimes come along. But if they do, they're unlikely to survive for long.

Picture on a world far, far away—or perhaps on the young Earth—that a microbe suddenly appears that's capable of absolutely accurate, foolproof self-copying. Elsewhere on the same world, other organisms are emerging whose self-replication machinery isn't quite so infallible. These slightly less-than-perfect self-copiers give rise to a genetically diverse population which natural selection gradually molds into a range of different species, each adapted to taking advantage of specific aspects of their increasingly complex environment. One very tempting aspect of this environment is a large reservoir of perfect, unchanging self-copiers—effectively, sitting ducks. In no time at all, natural selection comes up with a predator whose specialty is to attack and eat the perfect copiers which, being incapable of adapting and responding to the threat, are totally annihilated. (Of course, having used up its food source, the predator could quickly follow its prey into extinction.)

A perfect copier might seem like a good idea, but biologically it's a disaster, defenseless against changing conditions and more flexible competitors. The best kind of self-copier is one in which the copying mechanism is generally robust, but not invulnerable or completely flawless. It needs to be good enough to produce plenty of viable offspring that are themselves capable of reproduction. However, occasional copying mistakes are an absolute must. In genetics, variety is more than the spice of life—it's the essence of both survival and evolution.

Whether DNA and RNA exist elsewhere or not, the accurate synthesis of complex molecules and self-replication *with the possibility for occasional genetic change* demand substances of comparable status. To date, scientists know of no alternatives, except for hypothetical primitive precursors that would serve only in the early stages of life's development. It may well be that, like proteins and carbohydrates, the DNA/RNA partnership is simply the best arrangement available in the universe. If so, natural selection and convergence will ensure that it emerges repeatedly wherever biology takes hold.

✺

What will alien life look like? Most of it, almost certainly will be microscopic, as it is on Earth even today. We tend to forget that "advanced" life-forms like ourselves are the exceptions. To a good approximation, evolution produces microbes and little else. Your stomach alone is home to more bacteria than all the humans who've ever lived. Three-quarters of the bacterial species in your intestines have never been identified. In number of individuals, number of species, diversity of habitat, and length of tenure on this planet, microscopic organisms far outstrip their larger brethren, including animals and plants. From an astrobiological viewpoint, it may be most useful to think of all life, including Earth life, as various forms of microbial communities.

Our biased perspective leads us to see evolution as a steady progression from small to large, simple to complex, microbes to people, brainlessness to brains. But in many ways microorganisms have always been the dominant life-forms on Earth. Take away the smallest creatures and we'd be dead in no time—our very cells harbor vital components, such as the energy-generating mitochondria, that are essentially symbiotic bacteria. Take away us, on the other hand, and the bulk of terrestrial microbes would scarcely miss a beat.

Wherever there's life, there'll certainly be microbial life. On many worlds, life on the microscopic scale may be the *only* kind around. Bacteria and their kin almost certainly rule the galaxy in terms of sheer bulk. The jury is still out on how often the jump occurs to greater complexity, but Rare Earth scenarios notwithstanding, astrobiologists see no particular reason to assume that what has happened here is out of the ordinary.

Some biological developments appear to have no reasonable alternatives—among them, the early emergence of cells. Living things have to maintain an inner environment, including a collection of complex chemicals, that is radically different from their surroundings. The only way that's possible is by setting up a barrier—a wall or membrane—that encapsulates, circumscribes and protects the organism's contents. Since this feature of life is likely to be established almost from the start, it will almost certainly, given nature's parsimony, be incorporated into all future organisms as well.

The ubiquity of cells is suggested, too, by the fact that simple enclosing walls or membranes form remarkably easily in a variety of circumstances. As we saw in Chapter 3, cell-like structures emerge spontaneously when some organic materials found in meteorites are added to water. In fact, whenever oily or fatty chemicals occur in a watery environment they tend to arrange themselves into bubbles or globules that have the appearance of primitive cells.

Natural selection, acting on competitive protocells, will gradually refine their design and promote single-celled organisms of greater sophistication. On Earth, it resulted in microbes with a DNA-based genotype, a protein-based phenotype, and a lipid-based separator between the cell contents—the cytoplasm—and the outside world. While biologists can't be sure how many of the details of microscopic life will be the same on other worlds, they would be surprised if the differences were radical. All simple life, in essence, may be very similar. What about complex life? What, in particular, about multicellular life?

✿

It's easy to suppose that the jump from single-celled to multi-celled organisms was a unique and fairly recent event in Earth's history. But this isn't the case. The tendency of simpler creatures to come together in communities, which is at the heart of multicellularity, has shown itself many times and is almost as old as life itself. Early on, it led to the appearance, in shallow waters all over the planet, of microbial mats—masses of bacteria living close together and supporting a mutually beneficial food web. Cyanobacteria—a type of photosynthetic bacteria—living on the surface of these mats furnished food to microbes lower down that weren't equipped to harness light but that provided an anchor for the photosynthesizers. Sediment trapped between the layers of bacteria, or minerals deposited by them, accumulated as rounded stony pillars or stromatolites. Living stromatolite colonies still exist but only where conditions, such as high salinity, prevent them being attacked by predators. For a third of the Earth's history they were the dominant life-form on the planet, and fossils of them have been found dating back 3.5 billion years.

Apart from this well-publicized instance of communal living, bacteria are usually thought of as being strictly single-celled. Yet nothing could be further from the truth. Bacteria habitually form complex associations, hunt prey *en masse,* and lay down chemical trails to guide the movement of thousands of individuals. They specialize and routinely engage in intercellular communication. Many members of the group, known as Myxobacteria, never exist as isolated cells, even in their dormant state. As James Shapiro, a microbiologist at the University of Chicago explains:

> [T]hey enter dormancy in the form of a multicellular cyst that eventually germinates and spawns a ready-made population of thousands of individuals. Each cyst founds a new population; as the bacteria become more numerous and dense, a number of sophisticated events specific to multicellularity take place. Trails of extracellular slime are secreted and serve as highways for the directed movement of thousands of cells, rhythmic waves pulse through the entire population, streams of bacteria move to and from the center and edges of the spreading colony, and bacteria aggregate at specific places within the colony to construct cysts or, in some species, to form elaborate fruiting bodies.

In some cases, roaming colonies of Myxobacteria show multicellular responsiveness in identifying and attacking solitary prey microbes. They veer out of their way if they detect what might be food, and then turn to continue their search if the object proves inedible. Such purposeful, well-coordinated behavior is traditionally thought the preserve of larger organisms.

Its common existence at the microbial level suggests that the benefits of social living—of individuals forming cooperative assemblies—fuel a universal drive toward greater complexity and multicellularity.

We see evidence of this drive at other points in Earth's history. By the action of photosynthetic bacteria in microbial mats, our planet's atmosphere was changed from an anaerobic (oxygen-free) to an aerobic (oxygen-rich) state. Over a period of about 1.5 billion years, starting some 3.5 billion years ago, the oxygen content of Earth's atmosphere increased from next to nothing to a level of a few percent and continued to rise to its current level. Yet this increase wasn't steady; for more than a billion years after photosynthetic bacteria first appeared, the oxygen they produced didn't stay in the atmosphere, but instead reacted with iron and other minerals on land and dissolved in the sea. Only when the Earth's surface was fully oxidized did free oxygen start to build significantly in the air, resulting in the so-called Oxygen Revolution between 2.3 and 2.2 billion years ago.

From our viewpoint, this revolution was crucial. But oxygen is a double-edged sword. For many creatures that have evolved to live under oxygen-free conditions, it's a lethal gas, just as methane is to us. In fact, molecular oxygen is best thought of as a poison that attacks organic molecules, is deadly to the majority of anaerobic organisms (for example, *Clostridium tetani,* the bacterium that lives in deep puncture wounds and causes tetanus), and that forces major adaptations upon organisms that would be aerobic.

The Oxygen Revolution (or Oxygen Crisis, as it's also known) was pivotal in the rise of eukaryotes. These differ from prokaryotes (bacteria and archaea) in having much larger, more complex cells, including a nucleus where the genetic material resides, and various other membrane-bounded "organelles." Exactly how and when eukaryotes originated are subjects of ongoing controversy.

As atmospheric oxygen levels rose, some prokaryotic cells began to harness the immense power of oxygen metabolism to break down food sources into carbon dioxide and water. This new metabolic pathway yielded far more energy than any of the anaerobic pathways. If, as most biologists now believe, the endosymbiotic theory is correct, then some oxygen-metabolizing prokaryotes became permanently ensconced within early eukaryotic cells. The most popular idea is that they were originally ingested by eukaryotes as food, but managed to avoid being broken down. Subsequently, both parties benefited from the union. The aerobic incomers, which became mitochondria, found a sanctuary in which to safely live and breed, while their host cells acquired a high-energy, rechargeable battery pack. Plastids—the organelles responsible for photosynthesis in plants and algae—seem to have come about in the same way.

This symbiosis out of which modern eukaryotic cells emerged—cells living within cells—is another kind of multicellularity. Moreover, it made possible the rise of even more complex life-forms. The earliest eukaryotic fossils known, of a creature named *Grypania,* occur in chains up to 9 centimeters long and 1 millimeter in diameter—much too large to be single-celled. These fossils are about 2.1 billion years old, which places them in logical order close on the heels of the Oxygen Revolution. But in 1999, Jochen Brocks of the University of Sydney and his colleagues reported evidence of eukaryotes much older than *Grypania.* The evidence is in the form of organic molecules called steranes, detected in rocks 2.75 billion years old. Steranes can only come from the breakdown of complex sterols (a group of alcohols that includes cholesterol), and sterols are only made by eukaryotes using chemical pathways that demand molecular oxygen. A straightforward interpretation of these results is that eukaryotes were living 2.75 billion years ago with access to oxygen. Yet the Oxygen Revolution still lay 500 million years in the future. This presents paleobiologists with an interesting puzzle.

It also comes as bad news for a theory about eukaryotic origins put forward by Joseph Kirschvink, chief of Caltech's Paleomagnetics Laboratory, and which has been used as a Rare Earth argument. Kirschvink suggested a number of criteria that the prototype eukaryotic host cell had to meet. It had to be capable of phagocytosis (literally "cell eating", or ingesting food particles by surrounding them); be big enough to engulf other bacteria; and offer a controlled environment so that natural selection would favor it as a partner for symbiosis. Only one organism, he felt, met all the requirements: *Magnetobacter,* a Goliath among bacteria that uses onboard crystals of magnetite to orient itself along Earth's magnetic field lines. As we'll see in the next chapter, no such organism is likely to appear on an anaerobic world or on a world without a magnetic field. Kirschvink's proposal was used by Ward and Brownlee as one of their Rare Earth arguments because it implies special requirements for the rise of higher life.

If *Magnetobacter* were the ancestor of all eukaryotes, there ought to be no signs of eukaryotic activity before the Oxygen Revolution. Yet now we have evidence of eukaryotes living at a time when Earth's free oxygen levels must have been very low. It seems not only that complex life (at the level of eukaryotes) got started much earlier than was thought, but much earlier than it was thought possible.

A way out this enigma has been suggested by Harvard biologist Andrew Knoll. He takes as his central clue the fact that, together with the sterane biomarkers, Brocks and his colleagues found traces of what are called alpha-methylhopanes. These are only made by cyanobacteria—photosynthetic bacteria of the type found in microbial mats. It seems that oxygen-requiring eukaryotes were living in the same locale as oxygen-producing cyanobacteria. This is surely no coincidence. The stromatolites, Knoll suggests, probably provided an early, oxygen-rich oasis in which eukaryotes could get a head start. He counters possible criticism that oxygen released by stromatolites would be quickly diffused into the general ocean in three ways. First, many eukaryotic cells today can survive in oxygen-poor conditions, needing higher levels only in order to grow and reproduce. Second, the slime that microbial mats produce traps bubbles of oxygen, delaying the gas's escape. Third, the dependence of early eukaryotes on cyanobacteria would have encouraged close contact and made it easier for an endosymbiotic relationship to develop.

Who knows how much deeper in time we'll find evidence of eukaryotes? Just as scientists are continually pushing back the origin of life, we're learning of complex organisms and behavior in unexpectedly remote epochs—bacterial communities at least 3.5 billion years ago, nucleated cells less than 700 million years later. Life, it seems, was complex and had the propensity toward increasing complexity almost from the outset.

Multicellularity is not one of nature's recent inventions—an emergent quality that needed a long period of gestation. In the broadest and most meaningful sense, it's a general stratagem of life and one that affords multiple benefits. As Lynn Margulis and her son Dorian Sagan wrote, "Life did not take over the globe by combat, but by networking." There are many advantages to cooperation over going it alone. Cells within an ensemble can specialize, so that the collective can develop abilities far beyond the scope of a solitary, single-celled organism. A multicellular structure can be a more formidable predator and a less vulnerable prey—it's larger, relatively less exposed to its surroundings, and doesn't die just because one cell dies. It can adapt to colonize new environments and gain access to new resources—the ability to build a stalk, for instance, is a huge advantage for any organism that depends upon sunlight. More generally, multicellularity brings all the rewards and possibilities of being able to assume highly varied forms and structures, creating vast potential for evolutionary exploitation.

This morphological potential began to be fully realized during the Cambrian explosion, some 530 million years ago. Yet the significance of this event was not, as it's sometimes portrayed, the appearance of multi-celled organisms, but of *animals*. The Cambrian explosion represented just the latest and most dramatic exploration of the multicellular theme. Throughout the history of life on Earth, single-celled organisms have shown a tendency to progress at every opportunity toward some form of multicellularity. Microbial mats, tightly-knit bacterial colonies, individual eukaryotic cells, lichen (symbiotic associations of algae and fungi), colonial eukaryotes such as Volvox, as well as the more obvious examples of animals and plants, all display this trend for many cells to come together to form a cohesive and cooperative whole. Multicellularity can even be induced in the lab in populations that normally consist of single-celled organisms.

Martin Boraas and his colleagues at the University of Wisconsin in Milwaukee studied cultures of the green alga *Chlorella vulgaris*. The researchers had already shown that populations of the alga will remain single-celled for more than two decades, except for the occasional appearance of loose clusters of cells. But when they inoculated the cultures with a predatory single-celled flagellate, it was a different story. The algal population fell to begin with, but then recovered and was found to contain colonies made up of anything from four to hundreds of cells, as well as free-floating individuals. After a couple of months, there were hardly any single cells left and the bulk of the colonies were eight cells strong. The researchers observed that while flagellates could ingest single cells and young colonies, the mature colonies were too large for them to tackle.

When a feature of life crops up independently and persistently over time, it suggests a universal survival strategy at work. The evidence around us on Earth is that multicellularity is a ploy too good for nature to pass by—a convergent property that bestows such overwhelming advantages that we'll find it implemented routinely wherever living things emerge.

The same is true of mobility. Self-propelled organisms, as distinct from those that are sessile or at the mercy of natural currents, can elude predators, forage or prey on other creatures more effectively, and actively seek a more clement or fertile environment. The trade-off is that movement costs energy: a mobile organism has to be able to make up for its more profligate lifestyle by taking in more food. However, the strategy is evidently worthwhile, because many bacteria have their own form of locomotion. It consists of flagella—threadlike structures, one or more per organism, that protrude from the surface of the cells and can be spun around rapidly to produce movement. Many simple eukaryotes, like paramecium, also use flagella or cilia (numerous short flagella), though of a completely different design, for getting about. But with large multicellular eukaryotes, the evolutionary options opened up by morphological complexity have led to a much greater range of propulsion systems, from undulating bodies and fins to jointed limbs and vibrating wings.

Even intelligence appears to be convergent, and a more or less ubiquitous characteristic of life. As with multicellularity and motility, it's manifested to some degree in the simplest, most ancient creatures with whom we share the planet. That may sound far-fetched, until we adopt a less anthropocentric view of what intelligence entails. At the human level, we equate intelligence with a bundle of qualities, such as creativity, imagination, intuition, altruism, advanced problem-solving ability, language, and consciousness. But the most fundamental aspect of intelligence, in survival terms, is the ability to acquire, process and act upon information from the surroundings, an ability possessed even by prokaryotes. As James Shapiro explains, "The view that bacteria are sentient creatures, able to receive, process and respond meaningfully to external signals, has been gaining ground . . . as investigators spend more time exploring the mysteries of bacterial behavior."

The first inkling of microbial intelligence came in 1883, when the German biologist Wilhelm Pfeffer demonstrated that bacteria could analyze and compare stimuli. He filled capillary tubes with mixtures of repellants and attractants and showed that bacteria would swim into a tube to reach an attractant even if it first had to swim through a repellant—the equivalent of a human braving a hive of bees to get at the honey. Evidently, bacteria could make decisions, at least at a chemical level. But how?

Using strains of *Escherichia coli* whose pedigrees were known down to the gene, biochemists Julian Adler and his colleagues identified specific molecules that attract and repel the bacteria, and found that it isn't the quantity of stimuli to which the microbes are sensitive, but rather their *concentration gradient*—the increasing strength of a stimulus closer to its source. Daniel Koshland, of the University of California, Berkeley, while investigating this further, made a remarkable discovery: bacteria have a memory.

When *E. coli* aren't in a concentration gradient, they tumble randomly. But when they sense graded amounts of attractant, they immediately start using their flagella to swim smoothly on a straight course. Koshland wondered what would happen if bacteria were in a medium with an attractant mixed in. With no gradient they'd tumble. What if more attractant were suddenly and thoroughly mixed in? If cells analyze spatially, head-to-tail, they'd keep tumbling because, while there would be more attractant, there'd still be no gradient. On the other hand, if the cells *remember* the previous concentration, adding more attractant should trick them into thinking they're in a gradient, so that they start to swim stably. Koshland tried the experiment and, sure enough, mixing in more attractant fooled the bacteria; they began to swim in purposeful style.

There's also evidence that, as well as this short-term ability to recall concentrations, bacteria have a kind of long-term memory. Certain molecules act as sensory stimuli for all cells of a particular strain and can trigger behavior the first time the cell encounters them. Other molecules, however, become stimuli only if present while the cell is maturing. If a bacterium doesn't encounter these molecules when its young, it will never develop the mechanisms to perceive them.

Moving a rung or two up the ladder of complexity brings us to slime mold—again, not normally considered an intellectual heavyweight. Slime mold is a jelly-like creature that moves, as an amoeba does, by extending tube-like pseudopodia ("false legs"). Toshiyuki Nakagaki and his colleagues at the Bio-Mimetic Control Research Center in Nagoya, Japan, were interested to know how the slime mold Physarum polycephalum would cope with the challenge of a 30-square-centimeter agar gel maze comprising four possible routes. With no incentive on offer, the creature indiscriminately sent out a network of pseudopodia to fill all the available space. But when two pieces of food were placed at separate exit points in the labyrinth, the organism squeezed its entire body between the two nutrients along the shortest possible route, effectively solving the puzzle. The researchers believe the organism changed its shape to maximize its foraging efficiency and therefore its chances of survival. The meal of ground oat flakes led to a local increase in contraction of the organism's tube-like structures, propelling it toward the food. "This remarkable process of cellular computation," concluded the team, "implies that cellular materials can show a primitive intelligence."

Again, this isn't so surprising if we take a broad view of what counts as intelligent behavior. Even the humblest creature has to know and react to the difference between food and toxin if it's to survive. That involves not only receiving information but processing the information in order to be able to respond appropriately. Life and some level of intelligent behavior—discerning and doing what's best for one's survival—appear to go hand in hand.

It's obviously good to be able to outwit your competitors, seek food more effectively, and, in general, act in a way that promotes staying alive. Granted that behaving intelligently enhances survival, it's inevitable that natural selection will encourage its growth in some species. But increasing intelligence, like increasing motility, comes at a price. That becomes especially noticeable as we move forward in time and up the IQ scale to creatures with nervous systems and brains. Brains burn huge amounts of energy for their size: the human brain accounts for just 2 percent of body weight, but 25 percent of the body's caloric intake. Such extravagance has to be justified in terms of the edge that sophisticated intelligent behavior brings.

How common is intelligence at our level or higher? How often, across the galaxy and beyond, do big brains (or equivalent organs) arise? If four billion years is typical of the time it takes intelligence to evolve from scratch, it can be argued that it may be scarce. But as we saw in the last chapter, we've no way of knowing how the pace of evolutionary development on Earth compares with that in the universe at large. Even if it proves typical, many billions of stars in our galaxy alone are old enough to have allowed for the necessary incubation period. A recent analysis by Charles Lineweaver, an astronomer at the University of New South Wales, in Sydney, Australia, concludes that Earth-like planets around other stars will be, on average, 1.8 billion years older than Earth. His calculations take into account various factors that may determine the rate of formation and destruction of terrestrial-type worlds, such as the way in which heavy elements have become more plentiful since the Big Bang. As an aside, Lineweaver's results provide a plausible explanation for why attempts to detect signals from smart extraterrestrials have so far failed. Intelligent life on older worlds may have progressed as far beyond us as we have beyond bacteria; such beings are unlikely to communicate by any means as primitive as electromagnetic waves.

Another argument for high-level intelligence being rare is that it's rare on Earth. But it's important to avoid being overly anthropocentric on this point. Much of what makes humans seem to be on a different intellectual plane from other species on Earth, including our sophisticated language and advanced technology, is of recent origin and not due to any increase in raw brain capacity over the past few tens of thousands of years. Stripped of modern conveniences and our native tongue, we wouldn't appear to outshine some of our fellow terrestrials by quite such a wide margin. Many creatures, while not intellectual rivals of *Homo sapiens,* are not as far behind us as we sometimes suppose.

The expression "bird brain," for example, is inappropriate for describing any avian, but especially parrots and members of the Corvid family, which includes crows and ravens. Gavin Hunt, of Massey University in New Zealand, found that New Caledonian crows, living on a group of islands 1,400 kilometers northeast of Australia, use two distinct types of tools to forage for invertebrates such as insects, centipedes and larvae. Specialization in tool-making is something we tend to think is uniquely human. Yet Hunt saw the crows use a hooked tool made by plucking and stripping a barbed twig. He also observed the use of what he called a "stepped cut tool" with serrated edges, and found leaves from which crows had started to fashion these implements.

Ravens, larger cousins of the crow, excel at another talent we pride ourselves on—spontaneous problem solving. Bernd Heinrich of the University of Vermont raised five ravens under conditions that allowed him to know what learning experiences they'd been exposed to. He then tested their abilities to deal with a new situation by using pieces of meat hung by strings from perches. These strings were too long to let the birds reach down to grab the meat, and the birds were unable to capture the prize in mid-air by flying up to it, as the meat was too well secured. After many failed attempts, the birds began to ignore the food until, six hours into the experiment, one raven hit upon a solution. It reached down, pulled up as much string as it could, and trapped that length of string under its claws. Then it reached down again to grab some more and repeated the process until it had hauled the food up to its perch. Interestingly, there was no period of trial-and-error in this; the raven seemed to formulate a mental plan and carry it through successfully at the first attempt.

Several days later, a second raven solved the problem using a completely different method. In the end, four of the five birds independently arrived at different solutions. Only one failed—the same bird that also never learned that flying away with the tied-down meat in its beak always led to an unpleasant jerk when the food reached the end of its tether. Evidently, as with our species, some ravens are more bird-brained than others.

Irene Pepperberg, at the University of Arizona, has shown how adroit African gray parrots can be at using human language. Her star pupil, Alex, employs more than 100 English words to refer to all the objects in his lab environment that play a role in his life, including his 15 special foods, his gym and shower, and the experimenter's shoulder. At times he refuses to cooperate ("No!") and may tell the experimenter what to do ("Go away," "Go pick up the cup," "Come here.") He also requests particular information ("What's this?" "What color?" "You tell me.") After Alex had learned to use the numbers one through six and had grasped that a triangle is "three-cornered" and a square "four-cornered," he spontaneously and creatively called a pentagon "five-cornered." In formal, tightly controlled experiments, Alex is shown many objects in various combinations, and answers correctly an astonishing number of questions regarding these objects, such as "What object is red?" "What shape is [the object which is] wood?" "How many [are] wool?" And, "What color is the key?"

Our closest living relatives, the great apes, have shown an even greater proficiency with human language, and are clearly highly intelligent in other ways. The female lowland gorilla, Koko, for example, understands about 2,000 spoken English words, has a working vocabulary of over 1,000 signs (in American sign language) and is able to hold meaningful and interesting conversations with people. Dolphins and other toothed whales also display behavior that ranks them well up the intelligence scale.

Still, all birds and mammals are relative newcomers to the biosphere. The fact that we—the higher mammals, and human beings in particular—are here at all is thanks to a chance disaster that happened 65 million years ago. If the asteroid that struck the Earth and annihilated the dinosaurs had missed, mammals would probably still be mostly small, timid, nocturnal things cowering in the shadows of their saurian masters. It's a safe bet that if the Cretaceous extinction hadn't happened, *Homo sapiens* wouldn't be around. But does that mean there wouldn't have been advanced intelligence of some description? For almost 150 million years before the dinosaurs disappeared, there had been a slow but steady increase in the maximum ratio of brain to body mass. The smartest creatures, presumably, were getting smarter.

Some 75 million years ago there lived a small carnivorous dinosaur, about one and a half meters long and 40 kilograms in weight, known as Stenonychosaurus. From its remains, Dale Russell, curator of fossil vertebrates at the National Museums of Canada in Ottawa, estimated it had a brain-to-body mass ratio that would put it in the same intelligence league as an opposum. Nor was it alone; there were other comparably sharp-witted, small dinosaurs around at this time. At least as smart as their mammal contemporaries, these creatures were at the leading edge of a trend toward increased reptilian intelligence that might well have continued, had not a large rock from space handed the gauntlet to the mammals.

How far would reptilian brains have progressed? We don't know. But Russell offered his version of what might have happened if the descendants of Stenonychosaurus had gone on to develop intellectually at the same rate the most advanced mammals have done. To balance its heavy head, it would have had to stand upright. Russell postulated that a shoulder structure would have evolved to permit the "dinosauroid" to throw objects. Large eyes and three-fingered hands would have been legacies of its Cretaceous ancestor, he surmised, as would a lack of external sex organs—a general reptilian trait. At the same time, Russell gave his creature a navel, which is a novelty for a reptile but necessary if a placenta (present in some modern reptiles) evolved to help the birth of young with large crania.

The full-size model built according to Russell's design is an eerie meld of the familiar and the alien. In it, we see the convergionist's viewpoint fully and dramatically expressed. Rerun history without the Cretaceous extinction and Earth's masters would not be human. Yet high intelligence might still exist, encapsulated in this alternative and slightly disturbing form.

Does intelligence at our level or higher exist routinely in the galaxy? We cannot say, but it seems safe to suppose that worlds bearing very brainy creatures are going to prove much less common that worlds supporting life of any kind. Yet there may be hundreds of billions of habitable planets and moons in the galaxy. Even if only a thousandth of these gave rise to high intelligence, that would still be a staggering number.

✱

When we first look upon the biota of another world, we'll see no humans, elephants, butterflies, or roses. There will be genuinely alien life—different genetically, physiologically, anatomically, behaviorally. Yet in spite of the startling novelty, it won't all be unfamiliar. Eyes, ears, mouths, wings, legs, fins, leaves, roots, circulatory systems, guts, skin, scales, predators and prey, symbionts and parasites, instinct and intelligence, and many other aspects of the life we know will be in evidence, albeit in strange arrangements and guises. For wherever life emerges, it will have to evolve to deal with the same laws of nature and with the same kind of co-evolutionary pressures that we find on Earth. Convergence will ensure that we and our interstellar neighbors are, fundamentally, much more alike than we are dissimilar.

8
Life Signs

Confident that life inhabits many other worlds, astrobiologists are now pressing to find evidence of it. Life may take many forms and exist on many scales, from microscopic to planetary. It may dwell on or below the surface of a world, or in an ocean. It may be alive or extinct, within our solar system or on the planet of a star many light-years away. Across this wide spectrum of possibilities, astrobiologists are evolving the techniques needed to catch an uncertain and elusive prey.

What kind of life are they trying to find? By now, the answer should come as no surprise: Astrobiologists are looking for Earth-like life. This isn't because they're an uninspired bunch, unable to contemplate the wildly exotic, but because they increasingly believe that much of the life in the universe probably *is* Earth-like, at least in its basic biochemistry. If it isn't, they're in trouble, because the only life signs they have a reasonably clear idea of how to look for are those from broadly terrestrial-type organisms. Still, if bio-instruments did turn up anything out of the ordinary, they would at least sound an alarm. Then further research could be focused on finding out whether or not some unfamiliar biological process was at work.

☼

Two episodes have highlighted the problems scientists face in establishing the presence of extraterrestrial life. The life detection package carried aboard *Viking* in 1976 was the crowning achievement of the first fifteen years of what the geneticist Joshua Lederberg dubbed "exobiology." However, its legacy is troubling. One of the scientists involved, Norman Horowitz, concluded that *Viking* "found no life and found why there was no life," while another, Gilbert Levin, said "the scientific process forces me to my conclusion that there is microbial life on Mars." The other sobering experience has been the ongoing debate over the meaning of the purportedly biogenic traces in the Martian meteorite ALH 84001—a debate that remains unresolved despite several years of study in some of the most advanced laboratories on Earth.

Viking has taught us that metabolism is a tough criterion with which to send a robot probe looking for life. A couple of the experiments on the space-craft gave results that, by previously agreed criteria, were biologically posi-tive. Other data proved mystifying. It then fell to *Viking*'s mass spectrome-ter to cast the deciding vote of "no life," based on its failure to find organic material—a failure that itself left plenty of room for interpretation. There are too many ways in which ordinary chemical processes can mimic meta-bolic activity for this to be a reliable indicator. As Harvard biologist Andrew Knoll has pointed out, what you detect can't just be something life pro-duces—it must be something non-living things don't. An obvious principle, perhaps, but avoiding false positives is one of the most crucial aspects of the life detection game.

So what *should* astrobiologists look for, if not metabolism? Short of direct imagery of animals or plants, one of the clearest signs of life, as James Love-lock first explained back in the 1960s, is a planetary atmosphere that's way out of chemical equilibrium. According to Lovelock:

> It is a ubiquitous property of life to grow until the supply of materials sets a limit, and this applies to a whole planet just as much as to the growth of microorganisms in a test tube. Furthermore, an ecological steady state is established over the whole planet in which the surface and atmosphere are molded and changed to conditions which are optimum for the maintenance of the entire assembly of living creatures. Just as an animal is a cooperative assembly of living cells, so a planet can be considered as a living entity com-prising a cooperative assembly of species.

This, in a nutshell, is the Gaia hypothesis. According to Lovelock and the co-founder of the idea, Lynn Margulis, a biosphere can be thought of as a gigantic meta-organism that regulates its inner environment through a series of negative feedback linkages, in much the same way that ordinary creatures like ourselves keep our vital signs steady. James Kirchner, a geol-ogist at the University of California, Berkeley, has identified various flavors of the hypothesis, from weakest to strongest: Influential, Co-evolutionary, Homeostatic, Teleological and Optimizing. Take your pick, but you don't have to be a card-carrying Gaian to accept the truth of the least controversial form, which goes no further than asserting that biota have a big influence over certain aspects of the non-living world—the atmosphere being one of them.

Life on a large scale will push an atmosphere well out of chemical and thermodynamic kilter in ways that can be picked up over great distances. No alien scientist, having compiled a list of Earth's atmospheric ingredients, would have any qualms about stamping "PROBABLY ALIVE" right across it. In fact, measurements by the *Galileo* probe, as it swept by the Earth in 1990 to receive its final gravitational assist before heading off for Jupiter, provided reassuring support for the widely-held suspicion that life—even intelligent life—exists on our own world.

Galileo's confirmation came in the form of several independent readings. At near-infrared wavelengths (slightly longer than red light), the spacecraft recorded a strong dip in brightness at 0.76 micron, due to absorption by molecular oxygen. So pronounced was the feature that it told of an atmosphere many orders of magnitude richer in oxygen than any other place in the solar system. Immediately that suggests photosynthesis at work—but it doesn't prove it conclusively. As we saw earlier, a young Venus-like world in the throes of losing its water to ultraviolet photodissociation may have an oxygen-rich atmosphere, yet be utterly sterile.

Galileo's spectrometer also registered traces of methane—about one part per million—in Earth's air. Although that doesn't seem much, with such a lot of oxygen around any methane will be rapidly oxidized into carbon dioxide and water. Its presence in any quantity therefore implies that it's being constantly replenished, most likely by biogenic sources. In Earth's case, these include bacteria in bogs, in termites, and in those most agricultural of methanogens, flatulent cows and other ruminants.

Along with the spectral fingerprints of oxygen and methane, *Galileo* spotted a significant absorption band in the red part of the spectrum, around 0.7 micron, coming not from the atmosphere but from Earth's land masses. No known rocks or minerals give rise to such a feature. There is a very familiar substance, however, that absorbs red light in exactly this way and that can only be the product of living things: chlorophyll. Together with the presence of oxygen, it clinches the case for photosynthesizing life. Finally, with an instrument known as a plasma-wave detector, *Galileo* intercepted narrowband, pulsed AM radio signals from Earth, which confirm also another kind of life on the planet, brainy enough to have developed advanced technology.

It's comforting to know that our robot space probes have the capability to furnish strong evidence for life on the world of their creators. But, of course, the Earth has a huge and vibrant biosphere that extends thousands of meters below and above the entire surface. That certainly isn't the case on any other world in the solar system. No instruments, ground-based or space-borne, have ever detected anything about the Martian atmosphere to suggest it's other than abiotic in origin. The immediate implication is that, if there is presently life on Mars, it's very limited in scope.

This raises an interesting point: *can* life be so limited? Can it exist and survive indefinitely at a meager level, possibly in isolated pockets where locally favorable conditions, such as the occasional availability of liquid water, allow it to struggle along? A Gaian purist would say no. "Life is a phenomenon that exists on a planetary scale," Lovelock insists. "There can be no partial occupation of a planet by living organisms. Such a planet would be as impermanent as half an animal." It's a view that brooks no alternative— life is an all-or-nothing affair, and once it appears, it either embraces and transforms a world or it dies out. There's nothing in between. However, that's a mere article of faith without analytical support. Scientists at the cutting edge are working under the assumption that it isn't necessarily true. The consensus opinion in astrobiology is that life can be both rampant *and* fugitive, depending on the circumstances. In particular, it may start off being spread over a whole planet, then, if conditions turn more hostile, gradually retreat until it's reduced to scraping a living indefinitely in whatever sanctuaries remain.

Scientists still hope to find such refugee colonies on Mars. Those hopes were given a boost in 1999 by researchers at the University of Arkansas, who managed to grow bacteria on ash from a Hawaiian volcano—the closest earthly equivalent to Martian soil. Timothy Kral and his colleagues exposed the bacteria to a three-to-one mixture of hydrogen and carbon dioxide to simulate what they supposed conditions might be like up to three kilometers beneath the surface of Mars. All the bacteria survived, even when each gram of ash contained as little as 0.1 milliliters of water, by scavenging on sparse nutrients in the soil. In the process the bacteria released methane as a waste product.

Methane is one of the biological signs that researchers will be looking for on future automated missions to the Red Planet. It's already clear that the Martian atmosphere isn't significantly out of equilibrium; anything obvious would have been picked up by now. But even a miserly trace of methane would be enough to suggest that methane-producing microbes, similar to those in the Arkansas study, are alive somewhere beneath the surface.

Other instruments will be brought to bear above ground level on Mars. Having shied away from metabolism after the *Viking* experience, astrobiologists are now putting more faith in chemical biomarkers as signposts to life. One of these biomarkers is the anomalously high carbon-12 to carbon-13 isotope ratio that comes about when carbon is assimilated by living things. Such ratios, as we've already seen, have been used to provide the earliest (though not entirely undisputed) evidence of life on Earth. By equipping roving vehicles to drill samples of Martian rocks, and then examine their incinerated remains with a mass spectrometer, scientists will be able to seek out this biogenic C-12/C-13 ratio on the fourth planet.

Soon astrobiologists will have an impressive array of advanced biomarker detection gear at their disposal. Much of it is coming by way of other fields, ranging from medical technology and food safety to biological warfare defense. Although these instruments aren't all designed with space missions in mind, their miniature size and extraordinary sensitivity make them ideal for use in the search for life signs on Mars and elsewhere. Ideal, that is, except in one respect: the majority of this equipment, including an astonishing range of mass spectrometers, is designed quite narrowly to search for terrestrial biochemicals. Despite astrobiology's unavoidable slant toward Earth-centered biota, some instruments must adopt a less geocentric approach—a way of casting the net more widely while staying alert to the danger of false positives. With this in mind, the Jet Propulsion Laboratory issued, as one of its "Grand Challenges" to the science and engineering community, the task of coming up with novel strategies for life detection. Among the winners, and recipient of a grant to prove its concept, was a team from the University of Idaho.

The Idaho group is developing a device that, instead of looking for specific biomarkers, will test for life by trying to detect heat and energy-harvesting molecules produced by life processes. It's an approach that bridges the methodology gap between metabolism and biomarker tests by attempting to catch molecules in the metabolic act. At its most basic, life is about the chemical transformation of energy. In living systems, the transfer of energy from outside to inside involves a variety of electron-friendly molecules—electron carriers—that pass electrons like a bucket brigade from one to the next. In concert, these molecules serve as a multi-step, chemical power-pack.

The key electron transfer compounds for Earth life are quinones, porphyrins and flavins. Porphyrins, for example, form the basis of hemoglobin, which enables us to convey oxygen in our blood to our cells. A hint that life in other parts of the universe might use an Earth-related biochemical battery system comes from the discovery, mentioned in Chapter 3, that a type of quinone might be manufactured on the surface of interstellar ice grains. The great thing about the Idaho detector is its potential for finding any set of molecules that could form an electron transport chain, even one quite different from those we're familiar with. It will test for active life chemistry by checking for a variety of different charges on the candidate molecules and adding energy sources like mixtures of oxygen or iron and hydrogen to release energy.

Life detection becomes trickier if your quarry happens to be far underground. Then there may be no biogenic traces in the atmosphere at all, and surface clues might be inconclusive at best. Extreme subterranean biology is a real possibility on Mars, where a thick global layer of permafrost may overlie and isolate a deep hydrosphere that could be home to a vast community of rock-dwelling microbes. Testing this idea will be a major undertaking. It will involve drilling through as much as several kilometers of solid rock and ice in order to recover core samples for analysis—a task that only a well-equipped human expedition could attempt.

Recent evidence for liquid water near the surface of Mars, in the form of freshly cut channels, has once again revived prospects for life at shallower depths. But this doesn't mean the Martians will be easy to track down. The newfound channels—the intriguing "gully-washers" spotted from orbit by *Mars Global Surveyor*—are all in horrendously inaccessible places. Try designing and programming a spacecraft to land with pinpoint accuracy on the very rim of a crater or canyon, or worse, on the inner crater or canyon wall. NASA has had problems over the past couple of years just getting Mars landers down successfully on level ground.

Another key factor is the temporal question of whether we're dealing with existing or extinct life—a question that's particularly relevant in the case of Mars. Many scientists would stake their reputations, or at least a glass of beer, on life having developed or begun to develop on Mars in the dim past when it was warm and had plenty of surface water. Its continued presence today is touch and go.

One intriguing possibility is the reanimation of spores of Martian bacteria that have lain dormant for most of the planet's history. Scientists recently managed to revitalize spores of a now-extinct bacterial species that had been trapped within 250-million-year-old salt crystals recovered from a cave almost 600 meters underground in Carlsbad, New Mexico. This has prompted speculation that these tiny bodies are effectively immortal.

While we may or may not find living Martians, there's a terrific chance of coming across little fossils or other traces of ancient areobiology. And that brings us back to the enigmas contained within ALH 84001 and its sister fragments, the Nakhla and Shergotty stones. If the furor over the Martian meteorites has taught us anything, it's that very small shapes that look like fossils can be all sorts of things that have nothing to do with life. Round bacteria, or cocci, and spheres of inorganic origin such as gas bubbles are just about impossible for even an expert to tell apart. ALH 84001 has provided a valuable learning experience for astrobiologists, whatever the eventual verdict on its paleontological authenticity.

One deep truth that's emerged is that any evidence for or against extraterrestrial life based on minerals has to be eyed very suspiciously, for the simple reason that so much mineralization on Earth is biologically linked. To say whether samples from a fossilized Martian hot spring contain evidence of life, for example, it would be nice to know what a sterile hot spring looks like. But we don't have such a control. On Earth, every hydrothermal site is biologically prolific. What's been impressed upon astrobiologists over the past few years, especially as a result of the intense effort focused on the Martian rocks, is that life is fundamentally intertwined with geochemistry and geology.

Of the four main pieces of evidence from ALH 84001 cited in 1996 as indicating past life on Mars, every one has been found to have a plausible (though not necessarily correct) inorganic explanation. The lesson is that to establish the case for biominerals—minerals formed from long-dead organisms or as a result of their metabolic activity—you have to know about the wider conditions under which they were laid down. You have to know their provenance, geologically as well as biologically.

Sulfides are a good example. These are compounds formed when sulfur reacts directly with another element. Sulfides can be important indicators of the presence of life, *but only if they're laid down at low temperatures*. A hot origin would imply an ordinary inorganic process, but the only known way that sulfides can be made at cool (human-tolerable) temperatures is biologically.

In the case of ALH 84001 a big question is the temperature of the water that ran through the rock when the carbonate was deposited. Those who favor life say it was cool, while others argue it was very hot, maybe over 600°C, which is hopeless for anything organic. The clue we keep coming back to are those little magnetite crystals, which just might be the most important signs of extraterrestrial life found to date.

If you talk to Caltech's Joseph Kirschvink, he'll tell you a couple of very interesting things. The first is that the magnetite crystals in ALH 84001 have remained in perfect alignment since the day they were formed—a sure sign that they've never been at any temperature higher than 110°C. That's the upper limit—it could be that they've always been a lot cooler. In any event, we know of organisms that thrive above the boiling point of water. The second point is that there are six properties that, when found together, are diagnostic of biological magnetite. Some terrestrial magnetite of biological origin doesn't have all six properties; *all* terrestrial magnetite with all six properties is biological. And here's the clincher: some of the ALH 84001 magnetite has all six properties.

Is it a clincher, really? That depends on who you ask. Kirschvink is convinced, as are some of the original Johnson Space Center team who made the 1996 announcement. Others aren't so sure. A now-retired meteorite specialist at the University of California, San Diego, John Kerridge, points out that the magnetite crystals seem in some cases to have been formed, within the carbonates that hold them, through alteration. That would make them inorganic. So there's still room for doubt.

If, however, the magnetite from Mars is biological, then it opens up a variety of intriguing scenarios. One is that Mars and Earth developed differently. On Earth, evidence for magnetite-producing bacteria doesn't go back any further than about 2.4 billion years ago. That may be because these organisms need free oxygen, which wasn't plentiful until the Oxygen Revolution which occurred around this time. Another reason, again linked to the switch from a global anaerobic to aerobic regime, is that the ability to store iron, which Kirschvink argues provided a pre-adaptation to the making of biological magnets, isn't necessary when there's iron in solution all around, as there would have been in the pre-aerobic era. Now, the ALH 84001 carbonates are 3.9 billion years old. If there were biomagnetites on Mars at this time, perhaps there was oxygen, too. That would tie in with a suggestion by Christopher McKay and Hyman Hartman at NASA Ames that, with relatively little plate tectonics and vulcanism, Mars might have developed a biogenic oxygen atmosphere much earlier than Earth did by effectively burying lots of reduced organic carbon. The possibility of a primordial oxygen-rich atmosphere—not just on Mars, but on many worlds in the universe—opens up the possibility, McKay argues, of complex life becoming established much faster than it did on Earth.

The exhaustive working-through of these kinds of arguments has been one of the great benefits of the Martian fossils saga. It's led to the realization that we have to get a lot more systematic in identifying the unique characteristics of living things and the remains they leave behind. What biomarkers, if found in an extraterrestrial sample, would be accepted by the scientific community as a positive signal of life? That has to be established *before* we go out looking, not on the fly. As John Kerridge says, "It is important to decide which phenomena qualify as reliable biomarkers before using them to conclude that life was present. The ongoing controversy over . . . ALH 84001 . . . illustrates the danger of identifying certain phenomena as biomarkers at the same time as they are being used as evidence of life."

With a Mars sample–return mission on the horizon (albeit delayed), NASA is keen to have a set of appropriate biomarkers in place as soon as possible, and to that end set up a task force, "Biomarkers for Mars Exploration," in 2000. Similar issues are being addressed by the Life Detection Group, at JPL and Caltech, headed by Kenneth Nealson. The scramble is on to define, in analytical terms, life-as-we-know-it and the characteristic trails it leaves.

✷

If nothing else, the discovery of meteorites from Mars has shown beyond all doubt that material can be transferred, on a pretty routine basis, between one world of a planetary system and another. This raises another fascinating possibility. Could life hitch a ride on such impromptu interplanetary shuttles? There'd be ample shielding from harsh radiation inside a meteorite. Given the ability of some bacteria to survive almost indefinitely in a dormant state or as spores, even a flight time measured in millions of years wouldn't rule out organisms still being viable upon arrival at their new home. This possibility was given a further boost in October 2000 when Caltech's Benjamin Weiss and his colleagues (including Kirschvink) presented evidence that, during the journey of ALH 84001 from the surface of Mars to the surface of Earth, the interior of the meteorite was never heated to more than 40°C.

Presumably, just as asteroids have splashed bits of Mars our way, they've sent terrestrial emissaries in the other direction. What if creatures that originated on Earth ended up, billions of years ago, seeding the Red Planet with life? What if it happened the other way around, and we are descendants of ancient Martians? What if, in the remote past, there was a regular cross-fertilization of primitive life-forms between various worlds in the solar system, perhaps including the moons of the gas giants?

The possible biogenic nature of the remains in ALH 84001 adds a dash of immediacy to these speculations and compels astrobiologists to face a potentially frustrating issue. Even if they find indisputable evidence of past or present life on Mars, or anywhere else in the solar system, they won't be able to conclude right away that this is a totally independent instance of life. Think again about those magnetite crystals in ALH 84001. There are two possible reasons why, if they're biogenic, they look exactly like their terrestrial counterparts. One is that the magnetic bacteria which create these kind of structures are found all over the universe—a product of evolutionary convergence. The other is that they're directly related: the bacteria that make these magnets on Earth and on Mars ultimately had the same ancestor.

So, it isn't quite true to say that the top priority in astrobiology is to find the first example of extraterrestrial life. The top priority is to find the first example of life *that isn't part of the same phylogenetic tree*—that has no ancestral relationship whatsoever to life on Earth. If we find compelling evidence of life on Mars or any other world nearby, that will, of course, be tremendously exciting. But astrobiologists would immediately want to know if it was of entirely separate origin, or merely our distant relative by interplanetary inoculation.

Answering this question would be a lot simpler if we came across living Martian organisms. Comparing their genetic structure and DNA sequences with those of terrestrial bacteria would expose any family relationship. (If it turned out Martian life wasn't based on DNA or RNA, that would tell us in an instant it was completely alien.) It's also been suggested that we might look at the handedness of any Martian biological amino acids and sugars. If these differed from Earth life's rigid scheme of left-handed amino acids and right-handed sugars, it would demonstrate a separate origin—which is true. Unfortunately, the opposite is not true: if the handedness matched that of Earth biology it wouldn't necessarily prove a family link. For one thing, the odds of a match by chance alone aren't remote. For another, if handedness results from interstellar polarized light, you would expect the handedness of the biological molecules to be the same throughout the solar system, whether or not there'd been any cross-fertilization.

If astrobiologists find only signs of *past* life on Mars, or elsewhere close by, the question of its direct kinship with Earth life may be impossible to resolve. After all, the reason the claimed biomarkers in ALH 84001 were singled out in the first place is that they bear a striking resemblance to traces of terrestrial bacteria. It becomes difficult, after evidence for life is put forward based on similarity, to then turn around and use the same evidence to argue for life of a different kind. Is the Martian magnetite terrestrial in appearance because of evolutionary convergence or because of a common origin? Even if we were eventually to discover fossils and other indisputably biogenic traces on rocks examined *in situ* on Mars which look non-terrestrial, that wouldn't necessarily prove their genetic unrelatedness to us. Paleontologists continually find new kinds of fossils on Earth; biologists continually add to the number of known living species. So we couldn't say categorically that what appeared unique to Martian biology didn't in fact have a terrestrial cousin lying still undiscovered in the rock beneath our feet.

The other bind is that if primitive Martian biology—and we wouldn't expect it to be anything *other* than primitive—was so radically different from Earth life that it was obviously of separate descent, how would we identify it as being biological in the first place? By what criteria would scientists recognize its biogenic nature if it fell outside the previously-agreed morphological, chemical, isotopic and biomineralogical parameters for life? There are no easy answers. The way forward would be to carry out more research and assemble more data from various lines of inquiry, until a consensus emerged—exactly the approach, indeed, of any healthy, well-rounded branch of science.

Astrobiologists are coming to appreciate more and more that, barring something spectacular, like a large, obviously alien animal poking its nose into the camera of a future space probe, proving the existence of extraterrestrial life and its genetic distinctness isn't likely to happen as a result of any single experiment or mission. They will have to build a case, piecing together evidence from many different sources. Acceptance by the scientific community will probably be more like a grudging "Yes, I suppose it might be life" than a sudden "Eureka!"

❋

The proximity of Mars, its closest similarity to Earth of all the Sun's worlds, its intimations of water and a more clement past—these factors will keep drawing us back, again and again, until we know the truth: is there, or has there ever been, life on the fourth planet? The *Viking* mission, which raised more questions than it answered, was like a rite of passage for astrobiology. *Mars Pathfinder* was a brilliant success but of only marginal relevance to the search for life. *Mars Global Surveyor* has been another jewel in NASA's crown.

But, more recently, there's been the embarrassment of *Mars Climate Observer*—lost because one navigational team was working in imperial units and another in metric. Unfortunately, no one realized that until after the probe had fried in the Martian atmosphere. Close on the heels of this debacle came a second in the form of *Mars Polar Lander*—lost, basically, because the new NASA mantra of "faster, better, cheaper" had been favored over the older and wiser engineering dictum of "choose any two from three." Inadequate testing, overly ambitious design, and management mistakes brought about *MPL*'s downfall and, temporarily at least, that of NASA's reputation. However, lessons have been learned. NASA has reined in its short-term ambitions, moderated its "faster, better, cheaper" approach, and is ready to press on.

In 2001, there'll be a low-risk, confidence-building return to Mars in the form of a modest orbiter, equipped to tell us more about the mineralogy and surface chemical make-up of the planet, together with the kind of on-orbit radiation environment that future human travelers to the Red Planet can expect. If launched as scheduled on April 7, 2001, *Mars Odyssey* (named for the famous Kubrick/Clarke film and novel) will arrive in orbit on October 20. Then, in 2003, a more adventuresome campaign: two identical rovers, essentially scaled-up and more heavily instrumented versions of *Pathfinder Sojourner*, will land in different regions of Mars (still to be determined) using the same hit, bounce and roll airbag technology that got Pathfinder down safely. Unlike the little *Sojourner* rover, these meaner-looking explorers will be capable of traveling 100 meters in a day—as much as *Sojourner* managed in total. Yet their task will be similar—essentially, to gather data that's geological rather than biological. "The goal of both rovers will be to learn about ancient water and climate on Mars," explains Steven Squyres of Cornell University, the missions' principal science investigator. "You can think of each rover as a robotic field geologist, equipped to read the geologic record at its landing site and to learn what the conditions were like back when the rocks and soils there were formed." Since the history of climate and water on Mars is inextricably bound up with the question of Martian biota, the 2003 rover expeditions will play an important part in the "building a case for life" approach that astrobiologists are increasingly coming to adopt.

Also in 2003, a much smaller craft—smaller even than *Sojourner*—will arrive on the Martian surface having piggybacked a ride on the European *Mars Express* orbiter. *Beagle 2*, named after Darwin's ship, represents Britain's first independent interplanetary vehicle and a typical example of that country's penchant for doing top science on a shoe-string. More to the point, it's the *only* Mars lander firmly scheduled for the next several years that has a realistic chance of picking up life signs directly. Packed within its tiny total mass of less than 30 kilograms are instruments for probing both the atmosphere and surface for biological markers.

Having touched down on Isidis Planitia, a region of Mars believed to have been once inundated with water, *Beagle 2* will begin sniffing the air for trace gases that would strongly suggest a biogenic process at work. The key one of these gases is methane, the assumption being that if Mars has life at all it's overwhelmingly likely to consist of anaerobic bacteria that derive their energy from converting carbon dioxide into methane. A positive methane result would be hard to explain in any other way than through the action of near-surface-dwelling microorganisms. A negative result, though not bearing much on the possibility of life deep underground, would put tighter constraints on how much biological methanogenic activity could be going on near ground-level.

Another experiment on *Beagle 2* will look at the oxidative state of the surface. These results should help shed further light on the quarter-century-old mystery of the *Viking* data and whether they really can be explained in terms of highly reactive chemicals, as recent lab experiments suggest. Finally, and perhaps most intriguingly, the little craft will capture samples of rock, down to a depth of one meter, using a robotic mole equipped with modified dentist's drill. The samples will be subjected to "stepped" combustion—in other words, made to burn at different temperatures, from low to high. Then the incinerated, or oxidized, remains will be analyzed by a pocket-size mass spectrometer which will reveal the ratio of carbon-12 to carbon-13. An enrichment of carbon-12 would suggest a biogenic origin. The idea behind the *stepped* combustion is that organic compounds tend to vaporize at quite low temperatures—no more than a few hundred degrees—whereas, inorganic carbon-containing compounds burn at much higher temperatures. So this method allows a way of separating out any potentially organic stuff before putting it to the isotope test.

Looking further ahead, NASA has announced plans for a *Mars Reconnaissance Orbiter* in 2005, modeled on *Mars Global Surveyor* and capable of imaging objects as small as 30 centimeters (1 foot) across, and two missions in 2007 in collaboration with European countries. One of these will involve a large rover, plus possibly a balloon or airplane for low-level reconnaissance; the other will be a follow-on to *Mars Express* equipped with ground-penetrating radar to prospect for water. Finally, NASA hopes to launch a sample-return probe in 2011 that will bring back to Earth about 2 kilograms of Martian soil and rocks in 2014. Doubtless along the way, this schedule will change as new discoveries change the priorities for exploration.

In the end, Mars holds the answers but, like a good poker player, isn't being too forthcoming with clues. If there is life there, or ever has been, we're going to have to be persistent and creative in our efforts to find it. We may have to dig deep—kilometers deep. We may have to go to the polar caps and bore down to the base of the ice. And "we" may, literally, mean humans in the flesh. Perhaps the only way we're going to settle the issue of life on Mars that's dogged us for more than a century is to set up a human presence there with drilling gear, labs and all the paraphernalia that goes along with making life possible for a community of inquisitive but environmentally-finicky bipeds.

☀

Mars has shown how extraordinarily hard it is either to prove or disprove the presence of life even on our cosmic doorstep. Since the days of Schiaparelli and Lowell, it has tormented us with clues at the edge of resolution—clues that often turned out to be mirages even as they led us on to more tantalizing but never quite conclusive signs. We've even had bits of the planet lobbed our way, yet whether it's through telescopes, the eyes of visiting spacecraft or microscopes here on Earth, Mars continually manages to entice and tease us. Now we're experiencing *déjà vu* with Europa. Does it have a watery ocean or doesn't it? Is there life in the ocean or isn't there? Already we've been debating these questions for two decades.

Europa has been subject to a detailed survey by *Galileo,* which swooped past on several occasions during its peregrinations around Jupiter. The next step is the *Europa Orbiter* which will circle the moon exclusively and use radio wave echoes finally to settle the question of whether there's water beneath the icy crust and, if there is, at what depth. Originally scheduled for launch in 2003, the *Orbiter* has now been pushed back to 2005 or 2006 which, with a flight time of around five years, means that data won't be returned to Earth until the early 2010s. Fierce particle bombardment, courtesy of Jupiter's powerful radiation belts, is expected to knock out the *Orbiter*'s instruments after only a month or so, but hopefully not before the issue of Europa's long-debated ocean has been laid to rest.

After that, it's a question of budget and strategy. If mission planners go with the philosophy that they'll be able to design the follow-up, a *Europa*

Lander, most effectively once they've heard from the *Orbiter,* that would mean a *Lander* launch no sooner than 2015 and data back from the surface around 2020. Some scientists are arguing for an earlier departure on the ground that the choice of landing site isn't critical: we should just get there as quickly as we can and put down a probe to determine surface composition, salt content, acidity, presence of organics and so on. These are the clues we need to start figuring out what lies below.

Another interesting fast-track option, which was suggested and shelved but could easily be revived, is the *Europa Ice Clipper.* Using a radically new splash-and-grab approach to capturing samples for return to Earth, this could enable material to be brought back for laboratory analysis as early as the 2010s. In fact, it would make sense to run it as a dual mission with the *Orbiter* so that the latter could give navigational support. The *Ice Clipper* would zip over Europa and, shortly before closest approach, drop a hollow, 10-kilogram copper sphere that would smash into the moon's surface like a wrecking ball. The resulting plume of debris would then be intercepted by the spacecraft, and bits of it caught in an extraordinary substance known as aerogel (of which more below). Finally, *Ice Clipper* would swing around Jupiter and be hurled onto a path home. It's by far the fastest, cheapest and niftiest way of getting hold of fresh samples of Europa—even though they'd only be vaporized samples. Bringing back solid chunks of ice excavated from the moon's surface would push up the level of required technology and make for a bigger and more expensive mission.

Ultimately, of course, we want to put a probe *inside* the ocean—assuming there is one—to hunt for life directly. This poses an enormous challenge. Fortunately, we have a remarkable analogue of the supposed Europan sea right here on Earth where the techniques needed to explore beneath the surface of the Jovian moon can be developed and perfected. Almost 4,000 meters under the Antarctic ice, near the Russian Vostok Station, has been found a huge lake, probably of fresh water. Roughly the size of Lake Huron, measuring 14,000 square kilometers in area and over 500 meters in depth, and completely cut off from the surface, it represents the largest undisturbed, unexplored aquatic environment on the planet. A miniature remote-controlled submarine, known as a hydrobot, is being developed to enter it and search for any signs of life it might contain. One of the overriding concerns is to avoid contaminating Lake Vostok with substances or organisms from the outside world. What goes into this unique region must be totally sterile. Exactly the same care will be taken on Europa.

But how to insert a mini-sub below Europa's ice shell? This is where another prospective component of the life-seeking mission comes in. The idea, as it stands now, is for a long, pencil-shaped probe, with a heated tip— a cryobot—to melt its way down through the ice like a soldering iron through butter. Having done that, it will release two pieces of equipment. These will be set up where there seems the best chance of finding life—where energy gradients (so crucial to biology) are likely to be greatest. One, an ice–water interface station, will be sited just below the ice shelf so that it can monitor what happens there. The other, a sediment exploration station, will sink to the ocean floor and establish camp at the point of contact between the water at the seabed and the warmth welling up from the moon's interior. From this seabed station will emerge the hydrobot, to move and peer around, to explore any features of special interest, such as Europan hydrothermal vents, and generally seek out life anywhere from ocean floor to ceiling. Data will be relayed via the cryobot to equipment on the surface, and from there to Earth.

✳

The third of astrobiology's high priority targets in the solar system, after Mars and Europa, is Saturn's big moon Titan. It too will be the subject of intense investigation in the coming years.

Already en route to Saturn is *Cassini,* the last of NASA's old brigade of large, big-budget interplanetary probes. Launched in October 1997, *Cassini* is taking the scenic route, enjoying fuel-saving boosts from close encounters with Venus (twice) and Earth and Jupiter (once each), before arriving in orbit around the ringed planet in June 2004. The high point of the mission for astrobiologists will be the release of the European-built *Huygens* probe for a leisurely two-and-a-half-hour descent through Titan's dense, nitrogen-dominated atmosphere.

Having separated from its mother craft, the 2.7-meter-diameter *Huygens* probe will enter Titan's atmosphere three weeks later, on November 24, slowing down initially using its heat shield. Then at a height of about 175 kilometers, its main parachute will open, followed fifteen minutes later by a drogue chute. Throughout the remainder of its 140-kilometer descent, the probe will take measurements of the make-up, temperature, pressure, density and energy balance of Titan's atmosphere and relay this to the orbiter. Finally and most dramatically, it will send back the first clear pictures of the enigmatic moon's surface before touching down. What awaits it at journey's end, no one is sure—an ethane sea or a methane-ice island are high on the list of possibilities. Whatever lies in store, it will be the culmination of a great adventure: the first attempted landing on a world in the outer solar system, and the first in which we don't even know if the surface is solid or liquid. Having made land-or sea-fall, it's hoped that the probe will brave the elements long enough to continue transmitting data for up to half an hour.

Huygens isn't equipped to detect life, even in the remote event that any exists on Titan. But if all goes well, it will add greatly to the inventory of known chemicals in the moon's environment and, together with the *Cassini* orbiter, pave the way for a more ambitious mission, penciled in as the *Titan Biology Explorer*. This could take the form of a lander, a nuclear-powered helicopter, or, most likely, an aerobot—a steerable, balloon-borne craft that could carry out lengthy investigations of both atmosphere and surface, focused on uncovering the kind of prebiotic chemistry many researchers expect to find on this frigid but organically well-endowed world.

✻

What will we know about the possibilities for life elsewhere in the solar system by 2020? Although we're not likely to find life on Titan, we'll have learned how far prebiotic evolution can go in a place that has all the right ingredients but just happened to be in the wrong place. Titan will be our natural lab for understanding better the kind of processes that lead up to life.

Europa will probably still have us wondering. If it has an ocean, within twenty years we'll have confirmed its existence beyond all doubt. We may also have a pretty good idea of its depth and composition. But we won't yet have broken into it to see whether it harbors life. That great venture will probably come in the 2020s.

Mars is very hard to predict. If it stays true to form, it will still have us on the edge of our seats, giving out conflicting signals about its biological potential almost from one month to the next. Yet it's difficult to believe, given the intense focus of astrobiology on Mars, and the number of missions that will have probed and prodded the fourth planet in various ways over the next couple of decades, that we won't be a lot closer to a resolution. If we haven't found signs of past or present life by then, we'll have narrowed down severely where it might be. The final call may be left to human explorers, the first of whom will surely not arrive before about 2025.

✳

Of course, the Big Three targets don't exhaust the research goals that astro-biologists have within our planetary system. There's a lot more to be done close at hand than simply finding examples of life itself. The enigma sur-rounding the origin of life needs much deeper investigation. That will come not merely through more intense effort in labs on Earth, but through col-lecting samples from space of material that dates back to the very beginning of the solar system. Scientists want to know exactly what the solar nebula was like at the time when the planets were forming. One way to do this is to exam-ine meteorites that have freshly fallen—especially carbonaceous chondrites which might include virgin matter predating the Earth. That's exactly what researchers did in 2000 following the arrival and subsequent recovery of fragments of a large object, now known as the Tagish Lake meteorite, that came down in north-western Canada in January of that year. Subsequent analysis at Purdue University revealed it to be at least 4.5 billion years old and the first sizeable, uncontaminated chunk of preplanetary matter to come into science's possession.

Meanwhile, the *Stardust* probe, launched in 1999, is out roaming the solar system gathering up bits of dust that have been shed from comets or have breezed in on the interstellar wind. It uses aerogel, the lightest solid substance known, to slow down and capture fast-moving particles. Some-times referred to as "frozen smoke," aerogel is so light that a block of it the size of an adult person would weigh only 0.5 kilogram yet be strong enough to support the weight of a small car. In 2004, *Stardust* will encounter Comet Wild-2, analyzing its composition, taking pictures and harvesting more sam-ples before returning to Earth. Aboard a protective reentry capsule, the probe's collector panels will touch down, with the precision we've come to expect of space missions, in Utah in January 2006. Although we'll have to wait till then to learn the secrets of what has been caught in the aerogel trap, there's already cause for excitement. Mission scientists reported in 2000 that the probe had picked up some interstellar molecules of much greater mass than any previously known.

Comets are hugely important to our understanding of origins—of the solar system, of planets and of life itself. What is their detailed composition and structure? How far has the build-up of complex materials, and possibly even of prebiotic matter, progressed within them? To what extent have they influenced the origin, nature and development of life on Earth and on other worlds in space? At least three missions, scheduled for launch over the next few years, will help supply some of the answers by flying past, landing on and smashing into various cometary nuclei.

The first to depart, in July 2002, will be NASA's *CONTOUR*, destined to swoop past Comet Encke the following year and snap pictures of the nucleus from as close as 290 kilometers. Further encounters, with Comet Schwassmann-Wachmann-3 in 2006 and Comet d'Arrest in 2008, could be the prelude to a mission lasting up to thirty years.

In 2003, the European Space Agency probe *Rosetta* will blast off on a circuitous journey leading to an encounter with Comet Wirtanen eight years later. The culmination of a two-year observation phase will be the touchdown on the comet's nucleus of a multi-instrument lander to determine the surface composition to a depth of about a meter.

Exposing and analyzing material from further inside a cometary nucleus, however, demands more drastic action. NASA's *Deep Impact*, to be launched in 2004, will intercept Comet Tempel-2 in July of the following year and then release a 500-kilogram (approximately half-ton) copper cannonball. Smashing into Tempel's nucleus at 10 kilometers per second, this will excavate a crater some 120 meters wide and 25 meters deep, throwing out ice and rock fragments that will then be analyzed by instruments aboard the probe as it flies past them.

✲

Within the solar system we have the enormous advantage of being able to study worlds at close range. There's the thrilling prospect of actually getting our hands on an alien organism and bringing it back to Earth to study in the lab. But the drawbacks are serious: the difficulty of ruling out parochial effects, especially cross-fertilization, and worse, the limited number of targets, none of which offers a realistic prospect of advanced life.

Beyond lies the universe at large. Billions upon billions of worlds are out there, many of them surely teeming with life—and life, moreover, that will certainly be distinct in origin from our own. But their distances are terrifying. Our fastest robot probe would take hundreds of centuries to reach even the nearest star. Given that warp drive is still science fiction, we're compelled to squint for biological signs across tens or hundreds of light-years. If you were to have asked a scientist, fifty years ago, how to go about looking for life across interstellar distances, he or she would have probably said "Forget it. Wait till it comes to you."

Yet against all odds, astronomers are preparing to search for inhabited extrasolar worlds. Their efforts will revolutionize astrobiology, more so perhaps than spacecraft parachuting down out of the orange sky of Titan or roving the rock-strewn deserts of Mars. The world-shaking headlines of the next twenty years will likely come from giant instruments, on the ground and in Earth orbit, gazing with far sight at the planetary systems of other stars.

"We're now at a stage," explains extrasolar planet hunter Geoff Marcy, "where we are finding planets faster than we can investigate them and write up the results." For the first time, there's a planetary backlog—and it's getting longer by the month. We're on the brink of discovering much smaller worlds, and worlds in orbit at a comfortable distance from their central stars. Soon the hunt will begin in earnest for Earth-like worlds in Earth-like locations around stars resembling the Sun. Then, what is now a steady flow of discoveries will turn into a torrent.

Why this special interest in finding other planets like our own? The answer is that astrobiologists plan to use the weak Gaian strategy of looking for out-of-equilibrium gas compositions to detect extrasolar life. They're well aware that there could be exclusively sub-surface biospheres, for example, on the moons of gas giants. But the other-Earths approach offers the best chance of early success, because only surface life can be expected to have major effects on planetary atmospheres.

Already, several projects are underway that have the capability to pick up terrestrial-sized planets. Some of these make use of the fact that the gravitational fields of remote stars and any orbiting worlds they have can act like lenses. They can focus and distort the light coming from much brighter objects, such as galaxies, that lie at vastly greater distances behind them. The effect is called microlensing, and because it happens through chance alignments, there's no way of predicting when and where in the sky a microlensing "event" is going to take place. Astronomers get around this problem by simply watching the whole sky, and then zeroing in on a particular point if it seems that a microlens might have formed. At the first hint of a suspicious flickering or brightening, they immediately alert other researchers around the world who are part of the global microlensing network so that they can point their telescopes at the suspect's coordinates. Speed is essential because these things don't last long—just a few hours in some cases—nor do most of them involve planets. But a few have been recorded in which a brightening due to microlensing by an intervening star has an extra blip that suggests the presence of an orbiting world. One of these extra blips spotted in 1998 by a team at the University of Notre Dame has been attributed to a planet as lightweight as a few Earths orbiting approximately in its star's habitable zone—an exciting claim that still needs to be confirmed.

Microlensing studies may give some idea how commonly Earth-like planets occur in the galaxy at large. However, the objects they detect tend to lie at great distances, simply because of the way the lensing phenomenon works. What's more, the serendipitous nature of the method makes it difficult to apply to a thorough and detailed planetary survey.

Enter *Kepler*—a device in Earth orbit that will monitor, continuously and simultaneously, the light from 100,000 different stars in a small patch of sky in the constellation of Cygnus the Swan. Any tiny dip in brightness of a star could be due to a planet in transit—passing across the face of the star—as seen from our vantage point. The size and duration of such dips will be recorded by *Kepler,* and if they repeat exactly some time later will be taken as evidence of an orbiting planet whose size and distance from the star will be revealed by the details of the light changes. If given the go-ahead by NASA, *Kepler* could be launched on its four-year mission by 2005.

That will be an important first step. Using a large sample of stars, of all different types, *Kepler* will tell us how often Earth-sized planets are to be found circling in the habitable zones of their hosts, as well as many other details about planetary systems in general. But in the search for life, we need more specific information than this, and we also need to focus our attention in our stellar backyard. We need to identify Earth-like planets in the habitable zones of Sun-like stars that are relatively nearby, because these are the worlds most likely to have life that we can detect.

Scheduled for launch in 2006, the *Space Interferometry Mission* (SIM) will use a technique known as optical interferometry to sense wobbles in stars, up to 30 light-years away, caused by orbiting planets as small as the Earth. Optical interferometry works by combining the light collected by a number of smaller telescopes, all pointing at the same object, to achieve the resolution—the ability to distinguish detail—of a much larger single telescope. In SIM's case, two sets of four telescopes, each with a mirror 30-centimeters in diameter, will be arranged across a 10-meter boom to provide a resolution approaching that of a single 10-meter mirror. The result will be a tremendously powerful instrument when operated in the vacuum of space. But it still won't be powerful enough to look for the signatures of life. Having found nearby Earths, we have to gather up their light in sufficient quantities to be able to analyze their spectra for signs of gases like molecular oxygen, ozone and methane, that may tell us if biological processes are at work.

To give us the spectra of extrasolar Earths we'll need to put very large optical interferometers in space. One such instrument, under study by NASA, is the *Terrestrial Planet Finder* (TPF): an array of four 8-meter telescopes, with a combined surface area of 1,000 square meters, which could be flying as early as 2012. With such devices it will be possible to pick off, one by one, those worlds most like our own in the solar neighborhood that show the signatures of life. Then a new generation of devices will take over to glimpse the oceans and continents of these other Earths.

Picture five TPFs—each a mighty instrument in itself—flying in formation, spread out along a parabolic arc 6,000 kilometers in length. At the focal point of the array is a spacecraft that combines the light from the separate observing stations. This is the *Terrestrial Planet Imager* (TPI), capable of building pointilist portraits of nearby Earth-sized planets composed of 25 by 25 pixels or dots. That may not seem like the clearest of views. But remember, we're talking about planets the size of our own seen across voids of hundreds of trillions of kilometers. It is astonishing to think that we'll be able to glimpse the face of such worlds at all. And although the TPI will not show us the shorelines of islands or the meanderings of rivers, it will allow us to distinguish ice from water from land, and to look at the light from the land areas to see if it shows the characteristic absorption band of chlorophyll, in the same way that we see it from space on our own planet.

In the meantime, while these remarkable technologies are being designed, assembled and tested, there's plenty of theoretical work to be done. While we may have a good idea of the precise atmospheric features of a biosphere like that of today's Earth, the atmospheric and surface signatures of Earth's distant past are not fully understood. One of astrobiology's near-term goals will be to develop better global models for the Earth's early atmosphere, especially the pre-aerobic levels of biogenic gases like methane. We need to be able to discriminate between a geologically active but sterile planet (like an early Venus) and one whose atmosphere is under biological influence. We need to understand the coevolution of atmosphere and life, and how Gaia changes from birth to the present day.

✻

All these fabulous developments we can look forward to by 2020 or thereabouts. By then, we'll have a catalogue of tens of thousands of extrasolar planets that's growing by the day. We'll know what variety planetary systems can take and how unusual or typical is our own. We'll be aware of the number and location of Earth-like worlds within a radius of several tens of light-years, and we'll know which ones are the likely homes of life. That is an extraordinary prospect. Unless astrobiologists are very much mistaken, or we lose the willpower to carry through these great projects, within two decades we'll be able to point to some stars in the night sky and say "There live other creatures."

9

The Cosmic Community

Among a tiny community on the third planet of an unremarkable star in a typical galaxy in a universe some twelve billion years old, a new science has taken hold. What does this upstart field of astrobiology aspire to? What is its mission? Something quite extraordinary and profound: to grasp the history of life in a way that has not before been possible. Until now, we have had access to a single biological narrative—the complex yet parochial tale in which we ourselves are characters. How much of a part did chance play in the process that gave rise to *us*? To what extent could our story have been different? Terrestrial biology cannot answer these questions. But by revealing alternative sagas of life on other worlds, astrobiology will show which aspects of evolution are inevitable and which are capricious—contingent—wherever organisms appear.

Astrobiology will also let us see ourselves in a truer context. We've been alone to this point: the totality of known life confined to a single ball of rock in the vastness of space, and a common genetic heritage. Astrobiology will end that isolation. It will widen our sense of community to embrace the universe as a whole and all of its inhabitants. We'll begin to regard Earth not just as an ordinary planet but as an ordinary living world among many, and terrestrial life as but one species of a far more inclusive biota.

Ultimately, astrobiology aims for a unification as grand as any in particle physics: the unification of biology with cosmology. Its goal is nothing less than to understand, in a way that is both detailed and synthetic, how life springs forth from the evolution of the universe.

In the broadest sense, life everywhere enjoys a mutual kinship, a shared genealogy rooted throughout the cosmos in a common set of physical laws and raw materials. That much is already clear. But, additionally, astrobiologists have rallied around a number of ideas and principles that, based on our embryonic knowledge, seem most reasonable. These are the fundamental concepts informing our present-day searches:

1. Life is a universal phenomenon.
    Life is exclusively a planetary phenomenon (allowing for habitable moons).*
        Microbial life occurs commonly.
            Complex life (comparable to the level of animals) is not rare.
2. The most important defining characteristic of life is its ability to engage in Darwinian evolution.
    Life is a chemical system that can transfer its molecular information via self-replication and evolve via mutation and natural selection.
        Life is always based on carbon and water.
            Life is always based on proteins and nucleic acids.
        Life is always cellular.
3. Life originates on planets and moons.
    Abiogenesis (life from non-life) is "easy" and requires only concentrated organic chemicals, water and an energy source.
        Among the likeliest sites for abiogenesis are hydrothermal systems, especially deep-sea vents.
        The delivery of water and organic materials from space plays a significant part in the origin of life.
            Life may be transferred from one world to another, at least within the same planetary system.
4. Planets are very common.
    Habitable planets (and moons) are common.
        Planets and moons habitable on a long-term basis (billions of years) are not rare.
5. The evolution of life involves contingency and convergence.
    Of these, convergence is the more important.
6. Life can be both rampant (planet-wide) and refugial.

Such right now are our primary current expectations—the conjectures that, if proved correct, would least surprise astrobiologists. Yet there are bound to be surprises. Established fact in astrobiology remains a scarce commodity. This infant science is still at a stage where certain findings would

---

*The degree to which an item is indented in this list reflects how controversial it is.

send it careening along wildly divergent trajectories. In particular, if it turned out that evolutionary pathways were highly sensitive to initial conditions astrobiologists would have to return to the drawing board: most of those initial conditions are not yet known.

Throughout this book, we've focused on the emerging *consensus* in astrobiology—crucially, that it is beginning to seem more and more as if there's nothing special about what took place here on Earth. Life is widespread; life everywhere will have more fundamental similarities than differences; life will develop on a routine basis toward increasing complexity. These beliefs form the core of the current astrobiological paradigm. But let's end with some speculation. What would be the effect on the course of astrobiology if certain discoveries were made? Some of these hypothetical findings would fall more or less into line with present-day expectations; others would compel a deep reappraisal of our understanding of life in general.

1(a). Finding life on Mars

Martian organisms in general, past or present, would be one of the least unexpected discoveries in astrobiology and, in terms of what they could tell us about the universal properties of life, probably one of the least important. The consensus view is that all the essential pieces were in place for primitive life-forms to have emerged on Mars during its early clement phase some four billion years ago. Astrobiologists wouldn't be at all taken aback, therefore, to find biogenic traces in ancient Martian sediments or sites of hydrothermal activity. Nor would they be shocked to find extant microorganisms in locally favorable environments, either deep underground or nearer the surface where liquid water was still occasionally available.

Mounting evidence that simple life could survive a meteor-borne journey between Mars and Earth has dramatically cut the odds that interplanetary seeding took place, especially between 4.5 and 4 billion years ago. This is why finding life on Mars, although exciting, might not be as significant as other potential breakthroughs in astrobiology; most likely, the Martians would be our relatives. In fact, the more that Martian life was like Earth life, the less we'd stand to learn. Close similarities would almost certainly be a result of local (ballistic) panspermia.

If existing life turns up on Mars, its importance will be twofold. First, it will enable a detailed genetic comparison with similar terrestrial organisms to establish their degree of relatedness. Second, it will confirm what many astrobiologists suspect: that if necessity demands, life can survive indefinitely in a non–Gaian mode, exiled to certain regions of a planet where biological essentials remain in adequate supply.

1(b). A definitive finding that Mars has always been sterile

The more Martian life differs from Earth life, the more it would force astrobiologists to reassess their thinking. One major way it could differ is by being nonexistent. If life never appeared on Mars, even during its long-ago warm, wet phase, what would that mean? Life's rapid emergence on Earth has encouraged us to think that it's easy and not fussy about the initial conditions. It's absence on Mars would contest this view. A couple of possibilities would have to be considered, both of which imply that there are more constraints on abiogenesis than currently anticipated. First, it might be that there are one or more improbable steps in the process leading to life, so that even if there were a rerun of our own planet's dawn history, a biological outcome wouldn't be guaranteed. Second, it might be that something more specific to the primordial Martian environment wasn't conducive to life taking hold—something that by its presence blocked prebiological progress or by its absence meant that some crucial jump to increasing complexity couldn't take place. The discovery that Mars had always been sterile would also rule out Mars-to-Earth seeding as a possible explanation for the origin of terrestrial life.

2(a). A finding of life on Europa

Any biological discovery in the outer solar system would be thrilling because of the vanishingly small likelihood that it could have come about by interplanetary transfer. Whereas cross-fertilization is a distinct possibility in the case of Earth and Mars, the chances of living organisms or spores being transferred between our world and Europa, or vice versa, are negligible. Biology on Europa would almost certainly represent a truly independent instance of life. Because of this, it would immediately imply that life was common throughout the universe (unless it turned out that there was something unusual about the solar system as a whole).

Attention would focus on any similarities between Earth life and Europan because these would have to be taken as examples of convergence. For microbial life, we'd look at the basics: Is it cellular? Is it based on proteins and nucleic acids? If so, how closely do they resemble their terrestrial counterparts? Does Europan life depend on the citric acid cycle? How much diversity is there? Are there signs of incipient multicellularity, including endosymbiosis or microbial communities? Any features common to both terrestrial and Europan organisms would strongly suggest that these were universal attributes of life. Differences, conversely, would have to be entered on the other side of the ledger—items that bolster the case for contingency in evolution. At the same time, we would have to be careful about how the differences were interpreted. Europa's environment is not Earth-like, so that some biological idiosyncrasies will be due to local adaptations, not simply to a different roll of the evolutionary dice. That is to say, some non-terrestrial features of life *could* be universals that, for environmental reasons, didn't have the opportunity to find expression on our own planet. The only way astrobiologists could determine if this were the case—whether deviations from Earth-like biology were due to contingency or unfamiliar examples of convergence—would be by examining a range of different independent instances of life.

2(b). A finding of complex life on Europa

This would cause a sensation, and not just in astrobiology. Few scientists would bet on finding complex life, extant or extinct, anywhere else in the solar system. In the case of sub-ice oceans, it's thought unlikely that there would be a sufficient supply of organic chemicals or energy to support an advanced ecology. The disclosure of anything large and elaborate would take some explaining, rather like finding a Loch Ness Monster. How could complex creatures make a living in what appears to be an isolated and nutritionally poor environment—even allowing for a drip-feed of material from the surface through cracks and fissures?

Discovering a second instance of complex life would spectacularly advance the convergionist argument. It would prove what many astrobiologists suspect, that nature faces no special barrier in making the transition from microbial life to more sophisticated creatures. Yet "complex life" could mean many things. On Earth it includes animals, plants and fungi. If complex life is found on Europa or elsewhere, how many characteristics does it have in common with terrestrial varieties? Is it recognizably similar to any known non-microbial kingdom? Is it multicellular, motile, predatory? Does it show varied behavior, have nervous systems and brains? If it does, then all these features are confirmed as universals. And if they are, can intelligence, language, culture and technology be that much less likely? Finding anything resembling animal life on Europa wouldn't be quite the same as finding pointy-eared Vulcans, but they or someone like them could not be far behind.

3. A significantly out-of-equilibrium atmosphere on an extrasolar planet

In some ways this is the most important discovery that astrobiology can realistically hope to make in the next couple of decades or so. It would imply that life exists commonly as a *planetary* phenomenon—that (at least some of the time) it is not hidden away in the interstices of a world but dominates its geochemistry. Finding a second Gaia would show that there is nothing unusual about the rise of biology on a global scale. This level of complexity would be seen to be a normal outcome of planetary evolution.

Still, astrobiologists will need to be careful in interpreting data about out-of-equilibrium atmospheres. There are situations, as we saw in the case of a young Venus-like world, in which short-term chemical instability and the copious production of normally biogenic gases such as oxygen can stem from non-biological processes. Candidate Gaias would need to be (and surely would be) examined very closely for subtle combinations of atmospheric components that can't be explained, so far as we know, except as the result of life.

4. A deep, hot biosphere

This ranks among the more probable and far-reaching scenarios that we're considering here. Scientists already know of microscopic life on Earth inhabiting rock pores down to a depth of one or two kilometers. Cornell astronomer Thomas Gold has proposed that this "deep, hot biosphere" is much more extensive. In his view, the underground realm is where life originated, the *bulk* of terrestrial life is endolithic (rock-dwelling), and endoliths pervade the crust down to a depth of ten kilometers or more.

Gold's idea is that life may arise in any planetary interior where the combination of temperature and pressure allows water to exist as a liquid, and that such an environment is actually *more* conducive to life than planetary surfaces. He believes that "subsurface life may be widespread among the planetary bodies of our solar system, since many of them have equally suitable conditions below, while having totally inhospitable surfaces."

Ten worlds in the solar system, he suggests, may harbor deep, hot biospheres, including Mercury, Mars, Earth and the Moon—but not Venus, which lacks water. One key support for his model is the abundance of hydro-carbons in space; incorporated into planets and large satellites at their formation, these hydrocarbons then become the sources of chemical energy for life. According to Gold, petroleum and natural gas are not "fossil fuels"—they're not, as conventional wisdom insists, made from the remains of buried surface life. Instead, he argues, they well up from deep within the Earth, where they exist in virtually inexhaustible supply.

Verification of Gold's theory would open up the possibility of life not only on worlds normally considered biologically untenable, but also in virtually any planetary system. As he points out, "One may even speculate that such life may be widely disseminated in the universe, since planetary type bodies with similar sub-surface conditions may be common as solitary objects in space, as well as in other solar-type systems."

More than any other theory (including panspermia) Gold's deep, hot biosphere implies that planetary life is prolific across the universe. On a world where the surface milieu is permanently too hostile for life of any kind, the biosphere would remain entirely below ground level. On other worlds, like Earth and perhaps Mars, where surface conditions allow, some of the rock-dwelling life might migrate upward into the sunlight and evolve into more complex life-forms.

Of course, adaptation and colonization could work the other way. Life could first appear on a world near the surface, perhaps around hot springs or runoff pools, and then spread down into the crust as it adapts to living under more extreme conditions. Consequently, the discovery of a vibrant deep, hot biosphere on Earth wouldn't immediately bear out all of Gold's claims. We would still need to establish where life had first appeared and, further, whether it had also arisen under more or less the same conditions on some of our neighboring worlds. The deep, hot biosphere model can only be tested locally since there's no way, in the foreseeable future, to explore the interiors of extrasolar planets.

5. Bacteria in interstellar space

What if samples taken from comets, and from interstellar grains that have entered the solar system, are shown to harbor viable microbes? What if, in other words, the much-ridiculed theory of Fred Hoyle and Chandra Wick-ramasinghe turns out to be true?

Researchers would immediately want to know the nature of these space-faring organisms. Are they similar to terrestrial bacteria or viruses? Do they originate and evolve in the vacuum of space, or on the worlds of other stars? Astrobiologists would be keen to collect as many samples as possible in an effort to learn more about the structure, age and provenance of these unexpected creatures. They would also want to examine closely the surface of our neighboring planets and moons to see if the interlopers had established colonies anywhere else in the solar system. The best place to find pristine examples of biological material from interstellar space, as Carl Sagan pointed out, would be the moons of the outer planets—especially Neptune's biggest satellite, Triton.

The discovery of interstellar microbes or spores would suggest that star-to-star fertilization is possible and that panspermia might be a common way by which life is transferred to and established on infant worlds. It would also suggest that life is much more similar across the universe than might be the case if every planet were a separate site of abiogenesis. If terrestrial organisms have descended from living "seeds" that arrived here billions of years ago, astrobiologists would be faced with the questions of where life came from originally and how commonly it arises from scratch as opposed to being inseminated.

6. Artificial life

Some researchers, as we saw in Chapter 1, claim that "a-life" already exists on Earth in the form of computer-resident programs that replicate and evolve. This is a controversial view that challenges us to accept a much broader definition of life. Whether we're prepared to accept it or not, many researchers in computer science and robotics believe that machines will be developed during this century that look and act so lifelike that they will come to be considered synthetic organisms. Since technological evolution occurs several orders of magnitude faster than its biological counterpart, it's been conjectured that artificial life-forms, once established, will swiftly overtake organic life in many areas—including intelligence.

In a parallel development, we're likely to see an increasing union of human and machine. Already, small steps are being taken toward the integration of man and device in the form of electronic implants to improve hearing, sight, heart function, and mobility. Future innovations could include closer links between brains and computers to provide a way of artificially enhancing human intellect or connecting directly to the Internet.

Some SETI researchers believe that such developments are likely to be common among technological species. This has prompted speculation that, if and when we discover other intelligent life in the universe, much of it will be prove to be artificial.

A finding of artificial life, of course, immediately implies technology and therefore intelligence. Less obviously, it implies a community of intelligent species. While it is arguably unlikely that our galaxy contains just one intelligent species, it is certainly unlikely that it contains just *two*. (And if there are just two, both common sense and the classical ecological theory of "competitive exclusion" suggest that this number may soon be reduced to one.)

7. No life anywhere else in the universe

What if, after many years of searching, we find no trace of life at all beyond Earth? The implications would extend far beyond science into areas such as philosophy, metaphysics and religion. However, because the realization of the absence of other life would unfold very slowly—over many decades and even centuries—it would give us a long time to get used to the idea.

Inevitably, the discovery that we were alone would be taken by some as confirmation of the theological view that Earth and its inhabitants were divinely created. To many creationists, there's a major difference between the Rare Earth scenario and the Unique Earth scenario, where no life of any kind exists beyond our planet.

Even within science, however, at least one hypothesis predicts that we're unlikely to find life elsewhere. This is the cosmic anthropic principle, the gist of which is that the existence of life and, in particular, our presence as intelligent observers, severely constrains the nature of the universe. It was first discussed in 1961 by the Princeton physicist Robert Dicke, and has been developed by others, including Cambridge physicist Brandon Carter and one of the eminent pioneers of quantum mechanics, John Wheeler. In its weakest form, it simply points out that the universe has to be more or less the way it is or we couldn't be here. But some versions go much further, postulating that life is the product of such a remarkable chain of coincidences, extending from the Big Bang to contemporary Earth, that it may have arisen only once in the whole of cosmic history. Many astrobiologists, however, look at these ideas as philosophical musings that have little to do with observation and experiment.

8. Life with a fundamentally different basis

Suppose organisms are discovered whose biochemistry is completely alien to us. One consequence would be to undermine the first step of the convergionist argument—that life everywhere will tend to have a similar chemical basis, involving carbon macromolecules and water. However, this failure of the first step wouldn't necessarily have a knock-on effect causing the collapse of the convergionist case altogether; in fact, it could leave later steps in the argument unaffected. For example, if there exist large marine life-forms based on silicon, we should still expect them to have familiar anatomical adaptations for swimming—fins, streamlined shapes and so on—because the efficiency of such structures doesn't depend on the biochemistry in which they're implemented but on the laws of physics.

The greatest impact of finding life radically different from anything we know would be to expand the scope of astrobiology far beyond the mere discovery of a second instance of Earth-like biota. Just as uncovering one other example of carbon-and-water biology would suggest many more examples were waiting to be found, one case of life with a fundamentally different basis would show that our terrestrial notions of life were hopelessly parochial— the field would be opened wide to more extraordinary possibilities. Scientists would then face a double challenge: to come up with a much broader definition of life, and to develop ways of detecting organisms that had little in common with anything we had previously encountered. Many places deemed inhospitable to life would have to be reconsidered, and we would face the intriguing problem of recognizing life when we saw it.

9. Microbial life is common but complex life is rare

What if the "Rare Earth" scenario, discussed in Chapter 6, proves to be correct? Then the second step of the convergionist argument—the claim that multicellularity is universal—would be undermined. This could happen whether or not it turns out that life elsewhere can have a fundamentally different basis. Rare Earth implies that one or more stages involved in the evolution of complex, multicellular life are difficult, whatever the underlying biochemistry may be. What could cause that to be true? If complex life turned out to be very unusual, scientists would want to find out exactly what it was about the Earth that was so special. How could our own planet end up with so much "order for free," while the same natural laws operating elsewhere drew a blank? The apparent unreasonableness of this situation places the scenario of common life but scarce complexity among the strangest, least likely of those we have considered.

Astrobiology represents the final stage—and the final test—of the Copernican Revolution that began more than four centuries ago, when Renaissance astronomers plucked our world out of the center of the cosmos and revealed it to be a planet like any other. We stand on the threshold of a new era of thought and exploration. We've seen the notion that there's nothing special or privileged about our local circumstances fully vindicated in the physical sense. In just a few hundred years our home world has been transported from the center of all creation to a position of anonymity and cosmic insignificance.

Yet the effect of this dramatic shift hasn't been to make us *feel* insignificant but to place our history and circumstances within a pattern of expectation. The conditions that led to our present situation—the evolution of life, the evolution of the universe, Earth's history and local environment—are not a collection of unrelated anomalies explainable only as divine will but a coherent sequence of events, which we can understand and predict because we can observe its progress and its range of variation in countless places elsewhere.

Now we're about to test if the Copernican Revolution embraces fully the life sciences as well. Through the eyes of astrobiology we'll begin to appreciate how life on Earth fits into the scheme of life overall. That is an extraordinary prospect. The next two decades will see our view of the universe change beyond recognition. Within this period, many researchers feel confident, we'll uncover the first powerful evidence—possibly even proof—of extraterrestrial life. We'll begin to quantify the extent to which the cosmos is populated with habitable worlds, and perhaps gain some sense of life's total spectrum. For this reason alone, the opening years of the new millennium promise to be among the most enthralling of the human adventure.

Notes

Chapter 1. The Intimate Mystery

1   "The English geneticist J. B. S. Haldane . . ."
    Haldane, J. B. S. "What is life?" In his *Adventures of a Biologist*. New York: Macmillan (1937).

2   "Stanley Miller, a biochemist . . ."
    Quote from "Life: What exactly is it?" An Internet discussion among Stanley Miller, Antonio Lazcano, Mark Bedau, Thomas Ray, and Clair Fraser, moderated by Miranda Ran in *HMS Beagle: The BioMedNet Magazine*, issue 57 (June 25, 1999) (http://hmsbeagle.com/1999/57/viewpts/overview)

2   "Another origin-of-life researcher, Antonio Lazcano . . ."
    Ibid. See also: Lazcano, A., and Miller, S. L. "The origin and early evolution of life: Prebiotic chemistry, the pre-RNA world, and time." *Cell, 85*, 793–798 (1996).

2   "Mark Bedau, a philosopher . . ."
    Ibid. See also: Bedau, Mark. "The nature of life." In *The Philosophy of Artificial Life*, ed. Margaret A. Boden. Oxford: Oxford University Press (1996).

4   "Ray believes that systems . . ."
    Ray, T. S. "An evolutionary approach to synthetic biology: Zen and the art of creating life." In *Artificial Life 1*, pp. 195–226. Cambridge, Mass.: MIT Press (1994).

5   "Still, as one of the originators . . ."
    Langton, Christopher G. *Artificial Life: An Overview*. Cambridge, Mass.: MIT Press (1995).

5   "The naturalist D'Arcy Wentworth Thompson . . ."
    Thompson, D'Arcy Wentworth. *On Growth and Form, The Complete Revised Edition*. New York: Dover (1992). This second edition originally published in 1942 by Cambridge University Press. First edition published in 1917.

5 "More recently, the Chilean biologists . . ."
Maturana, Humberto R., and Varela, Francisco J. *Autopoiesis and Cognition: The Realization of the Living*, Boston Studies in the Philosophy of Science, vol. 42. Dordrecht: D. Reidel (1980).

5 "As Stuart Kauffman, at the Santa Fe Institute . . ."
Kauffman, Stuart. *At Home in the Universe: The Search for the Laws of Self-Organization and Complexity*. New York: Oxford University Press (1995).

6 "When you think about . . ."
Michael Meyer quote from Holmes, R. "Life is . . ." *New Scientist*, p. 40 (June 13, 1998).

8 "Could life exist without . . ."
Margulis, Lynn, and Sagan, Dorian. *What is Life?* New York: Simon & Schuster (1996).

8 "Jeffrey Bada, who directs . . ."
Bada quote from Cohen, P. "Let there be life." *New Scientist*, pp. 22–27 (July 6, 1996).

9 "'The ability to evolve,' insists Jack Szostak . . ."
Quoted by Robert S. Boyd, "In labs, search for how life began." *Philadelphia Inquirer* (January 2, 1999).

9 "As Oxford biologist John Maynard Smith . . ."
Maynard Smith, J. *The Problems of Biology*. Oxford: Oxford University Press (1986).

9 "The Oxford biologist Richard Dawkins . . ."
Dawkins, Richard. *The Selfish Gene*. New York: Oxford University Press (1978).

10 "A more conventional view of life . . ."
Hull, D. L. "Units of evolution: A metaphysical essay." In *The Philosophy of Evolution*, ed. U. J. Jensen and R. Harré, pp. 23–44. New York: St. Martin's (1981).

10 "According to Carl Emmeche, a philosopher . . ."
Emmeche, C. "Autopoietic systems, replicators, and the search for a meaningful biologic definition of life." *Ultimate Reality and Meaning*, 20(4), 244–264 (1997).

12 "The originator of the Gaia theory . . ."
Lovelock, James. *The Ages of Gaia*. New York: W. W. Norton (1988).
Lovelock, James, and Epton, Sidney. "The quest for Gaia." *New Scientist* (February 6, 1975).

12 "In The Black Cloud . . ."
Hoyle, Fred. *The Black Cloud*. New York: Signet (1959)

12 "Aerospace engineer Robert Forward . . ."
Forward, Robert L. *Dragon's Egg*. New York: Del Ray (1980).

13   "It's evident in the 'Roadmap' . . ."
       DeVincenzi, D., ed. "Final report." Astrobiology Workshop at NASA
       Ames Research Center (December 1996).

13   "'No other element comes close . . .'"
       Bada quote in Holmes, R. "Life is . . ." *New Scientist*, p. 40 (June 13, 1998).

13   "In 1893, the chemist James Emerson Reynolds . . ."
       Reynolds, J. E. *Nature, 48*, 477 (1883).

13   "Picking up on this idea . . ."
       Wells, H. G. "Another basis for life." *Saturday Review*, p. 676 (December
       22, 1894).

14   "This difficulty didn't faze Stanley Weinbaum . . ."
       Weinbaum, Stanley G. *A Martian Odyssey.* New York: Fantasy (1949).

14   "He said: 'I tell my students . . .' "
       Recounted by Simon Conway Morris in his presentation "Evolution
       bound: The ubiquity of convergence" at the First Astrobiology Science
       Conference, NASA Ames (April 2000).

14   "Harold Morowitz, a biologist . . ."
       Morowitz quote in Holmes, R. "Life is . . ." *New Scientist*, p. 40 (June 13, 1998).

14   "Others, like Christopher Chyba . . ."
       Chyba, C., and McDonald, G. "The origin of life in the solar system: Cur-
       rent issues." *Annual Review of Earth and Planetary Science, 24*, 215–249 (1995).

Chapter 2. Original Thoughts

17   "In 1924, the Russian biochemist Alexander Oparin . . ."
       Oparin, A. I. *The Origin of Life.* 2nd ed. New York: Dover (1953). Russian
       edition published in 1924. First English edition published in 1936.

17   "Four years later, and quite independently . . ."
       Haldane, J. B. S. "The origin of life." In *Rationalist Annual* (1929).
       Reprinted in J. B. S. Haldane, The Inequality of Man. London: Penguin
       (1937).

17   "In a classic paper, called 'The Physical Basis of Life' . . ."
       Bernal, J. D. *The Physical Basis of Life.* London: Routledge and Kegan Paul
       (1951).

18   "In 1929, in *The World, the Flesh, and the Devil* . . ."
       Bernal, J. D. *The World, the Flesh, and the Devil: An Enquiry into the Future
       of the Three Enemies of the Rational Soul.* 2nd ed. Bloomington: Indiana
       University Press (1969). First published 1929.

18   "Half a world away, in Chicago . . ."
       Miller, S. L. "A production of amino acids under possible primitive Earth
       conditions." *Science, 117*, 528–529 (1953).

19   "During the question period that followed . . ."
        Arnold, James R., Bigelstein, Jacob, and Hutchison, Clyde A., Jr. *Harold Clayton Urey*, Biographical Memoirs, vol. 68, National Academy of Sciences (1996).

21   "The first person to look upon this astonishing menagerie . . ."
        Corliss, J. B., Baross, J. A., and Hoffman, S. E. "An hypothesis concerning the relationship between submarine hot springs and the origin of life." *Oceanologica Acta, 4,* suppl. C4, 59–69 (1981).

22   "Heat-loving bacteria had already been found . . ."
        Brock, T. D., and H. Freeze. "Thermus aquaticus gen. n. and sp. n., a non-sporulating extreme thermophile." *Journal of Bacteriology, 98,* 289–297 (1969).

23   "June 2000 brought news that Birger Rasmussen . . ."
        Rasmussen, B. "Filamentous microfossils in a 3,235-million-year-old volcanogenic massive sulphide deposit." *Nature, 405,* 676–679 (2000).

24   "Their remains were dug up in 1993 . . ."
        Schopf, J. W., and Procter, B. M. "Early Archaean (3.3 billion to 3.5 billion-year-old) microfossils from Warrawoona Group, Western Australia." *Science, 237,* 70–73 (1987).

24   "During a 1991 expedition . . ."
        Mojzsis, S. J., Arrhenius, G., McKeegan, K. D., Harrison, T. M., Nutman, A. P., and Friend, C. R. L, "Evidence for life on Earth before 3,800 million years ago." *Nature, 384,* 55–59 (1996).

25   "'There is confusion . . .'"
        Quote by Mojzsis from a private communication.

26   "The ancient Greenland rocks . . ."
        Pointed out by Gustaf Arrhenius in a private communication.

26   "The point the dissenters are making . . ."
        Lepland, A., and Arrhenius, G. "Accretion of moon and Earth and the emergence of life." *Chemical Geology, 169,* 69–83 (2000).

27   "It could have happened if the Earth . . ."
        Canup, R. M., and Esposito, L. W. "Accretion of the moon from an impact-generated disk." *Icarus, 119,* 427–446 (1996).

28   "A different model for the Earth's formation . . ."
        Turekian, K. K., and Clark, S. P. "Inhomogeneous accumulation of the Earth from the primitive solar nebula." *Earth and Planetary Science Letters, 6,* 346–348 (1969).

28   "Could life have started well before four billion years ago?"
        Wilde, S. A., Valley, J. W., and Peck, W. H. "Evidence from detrital zircons for the existence of continental crust and oceans on the Earth 4.4 Gyr ago," *Nature, 409,* 175–178 (2001).

29   "Wächtershäuser is a German scientist . . ."

Wächtershäuser, G. "Evolution of first metabolic cycles." *Proceedings of the National Academy of Sciences USA, 87,* 200–204 (1990).

Wächtershäuser, G. "Pyrite formation, the first energy source for life: A hypothesis." *Systematic and Applied Microbiology, 10,* 207–210 (1988).

30   "In this way, Carnegie researchers have managed to make ammonia . . ."

Brandes, J. A., et al. "Abiotic nitrogen reduction on the early Earth." *Nature, 395,* 365–367 (1998).

30   "At about the same time, early in 1999, Koichiro Matsuno . . ."

Imai, E., et al. "Elongation of oligopeptides in a simulated submarine hydrothermal system." *Science, 283,* 831–833 (1999).

31   "At one extreme of the environmental spectrum . . ."

Gold, Thomas. *The Deep Hot Biosphere.* New York: Springer-Verlag (1999).

31   "At the other extreme, intriguing new evidence . . ."

Dobson, C. M., Ellison, G. B., Tuck, A. F., and Vaida, V. "Atmospheric aerosols as prebiotic chemical reactors." *Proceedings of the National Academy of Sciences, 97,* 11864–11868 (2000).

Chapter 3. Star Seed

33   "As early as the first half of the nineteenth century . . ."

Berzelius, J. J. *Annalen der physikalisches Chemie, 33,* 113 (1834). Reported finding humic acid in the Alais carbonaceous chondrite.

34   "Researchers had reported finding amino acids in meteorites . . ."

Mueller, G. "Amino acids in the Cold Bokkeveld carbonaceous chondrite". *Geochimica et Cosmochimica Acta, 41,* 1325 (1953).

35   "Furthermore, all the Murchison amino acids . . ."

Kvenvolden, K. A., et al. *Nature, 228,* 923–926 (1970). Report that amino acids and other organic molecules in Murchison meteorite are extraterrestrial because they are racemic.

35   "In 1982, geochemists Michael Engel and Bart Nagy . . ."

Engel, M. H., and Nagy, B. "Distribution and enantiomeric composition of amino acids in the Murchison meteorite." *Nature, 296,* 837–840 (1982).

35   "Even though Engel and Stephen Macko . . ."

Engel, M. H., Macko, S. A., and Silfer, J. A. "Carbon isotope composition of individual amino acids in the Murchison meteorite." *Nature, 348,* 47–49 (1990).

Engel, M. H., and Macko, S. A. "Isotopic evidence for extraterrestrial nonracemic amino acids in the Murchison meteorite." *Nature, 389,* 265–268 (1997).

35   "In 1997, chemists John Cronin and Sandra Pizzarello . . ."

Cronin, J. R., and Pizzarello, S. "Enantiomeric excesses in meteorite amino acids." *Science, 275,* 951–955 (1997).

36   "One intriguing source was proposed . . ."
      Bonner W. A., and Rubenstein E. *BioSystems, 20*, 99–111 (1987).

36   "James Hough at the University of Hertfordshire . . ."
      Bailey, J., et al. "Circular polarization in star-formation regions: Implications for biomolecular homochirality." *Science, 281*, 672–674 (1988).

37   "A possible solution to the second problem . . ."
      Shibata, T., et al. "Amplification of a slight enantiomeric imbalance in molecules based on asymmetric autocatalysis: The first correlation between high enantiomeric enrichment in a chiral molecule and circularly polarized light." *Journal of the American Chemical Society, 120*(46), 12157–12158 (1998).

37   "But as James Hough pointed out . . ."
      Guterman, Lila. "Why life on Earth leans to the left." *New Scientist*, p. 16 (December 12, 1998).

40   "David Deamer, a chemist . . ."
      Deamer, D. W., and Pashley, R. M. "Amphiphilic components of the Murchison carbonaceous chrondite: Surface properties and membrane formation." *Origins of Life and Evolution of the Biosphere, 19*, 21–38 (1989).

41   "Measurements of the precise wavelengths and structure of the DIBs . . ."
      Desert, F.-X., Jenniskens, P., and Dennefeld, M. "Diffuse interstellar bands and UV extinction curves: The missing link." *Astronomy and Astrophysics, 303*, 223–232 (1995).

41   "To investigate what might happen next, Bernstein . . ."
      Bernstein, M., et al. "UV irradiation of polycyclic aromatic hydrocarbons in ices: Production of alcohols, quinones and ethers." *Science, 283*, 1135–1138 (1999).

43   "This idea took a knock, however, in March 1999 . . ."
      Blake, G., et al. "Sublimation from icy jets as a probe of the interstellar volatile content." *Nature, 398*, 213–216 (1999).

44   "As the University of Hawaii astronomer Tobias Owen put it . . ."
      Owen, Tobias. Scientific American "Ask the Expert" web page http://www.sciam.com/askexpert/environment/environment13.html

44   "But then he, Scripps colleague Luann Becker . . ."
      Bada quote from Cohen P. "Let there be life!" *New Scientist*, pp. 22–27 (July 6, 1996).

44   "Prospecting around this structure in 1994, Bada and his colleagues . . ."
      Becker, L., Bada, J. L., Winans, R. E., Hunt, J. E., Bunch, T. E., and French, B. M. "Fullerenes in the 1.85 billion-year-old Sudbury impact structure." *Science, 265*, 642–645 (1994).

44   "Closer study, though . . ."
      Becker, L., Poreda, R. J., and Bada, J. L. "Extraterrestrial helium trapped in fullerenes in the Sudbury impact structure." *Science, 272*, 249–252 (1996).

45   "Most dramatically, in March 2000, Becker, Poreda, and Ted Bunch . . ."
       Becker, L., Poreda, R. J., and Bunch, T. E. "Fullerenes: An extraterrestrial
       carbon carrier phase for noble gases." *Proceedings of the National Academy
       of Sciences USA, 97*(7), 2979–2983 (2000).

45   "In 1999, Elisabetta Pierazzo of the University of Arizona and Chris
       Chyba . . ."
       Pierazzo, E., and Chyba, C. F. "Amino acid survival in large cometary
       impacts." *Meteoritics and Planetary Science, 34*(6), 909–918 (1999).

45   "As Bernstein, Sandford, and Allamandola have said . . ."
       Bernstein, Max P., Sandford, Scott A., and Allamandola, Louis, J. "Life's
       far-flung materials." *Scientific American, 281*(1), 42–49 (July 1999).

46   "Bernstein, Sandford, and Allamandola again . . ."
       Ibid.

46   "Since the 1970s, Hoyle and his Sri Lankan-born colleague . . ."
       Hoyle, F., and Wickramasinghe, C. "Prebiotic molecules and interstellar
       grains." *Nature, 266*, 241 (1977).

47   "Most extraordinary of all, Hoyle and Wickramasinghe . . ."
       Hoyle, Fred, and Wickramasinghe, Chandra. *Evolution from Space.* Lon-
       don: J. M. Dent & Sons (1981).

47   "Svante Arrhenius, grandfather of Scripps oceanographer . . ."
       Arrhenius, S. "Panspermy: The transmission of life from star to star." *Sci-
       entific American, 96*, 196 (1907).

48   "According to *Apollo 12* commander Pete Conrad . . ."
       Conrad's comments on the bacteria found on *Surveyor 3* camera recorded
       in the *Apollo 12 Lunar Surface Journal,* ed. Eric Jones. Washington, D.C.:
       NASA (1995).

48   "Murray's idea has recently received experimental support . . ."
       Battista, J. R., Earl, A. M., and Park, M-J. "Why is Deinococcus radiodu-
       rans so resistant to ionizing radiation?" *Trends in Microbiology, 7,* 362–365
       (1999).

49   "Among the surprises: samples of Bacteria subtilis . . ."
       Horneck, G. "Exobiological experiments in Earth Orbit." *Advances in
       Space Research, 22*(9), 317–326 (1998).

49   "Gerda Horneck, a specialist in radiation biology . . ."
       Horneck, G. "Panspermia revisited: Can microbes survive a trip through
       the solar system?" Oral presentation at the First Astrobiology Science Con-
       ference, NASA Ames (April 3–5, 2000).

49   "Stanford geologist Norman Sleep has calculated . . ."
       Zahnle, K. J., and Sleep, N. H. "Impacts and the early evolution of life."
       Comets and the Origin and Evolution of Life, ed. P. J. Thomas, C. F. Chyba,
       and C. P. McKay, pp. 175–208. New York: Springer-Verlag (1997).

50   "The idea put forward by astrobiologist Kevin Zahnle . . ."

Ibid.

50   "In the case of the solar system, as Zahnle points out . . ."
Zahnle, K. "Life and death on Mars." Poster presentation at the First
Astrobiology Science Conference, NASA Ames (April 3–5, 2000).

Chapter 4. Havens, Hells, and H₂O

54   "Water seems ordinary, but it isn't . . ."
Ball, Philip. *Life's Matrix: A Biography of Water.* New York: Farrar, Straus
and Giroux (2000).

55   "Here, for instance, is the astronomer James Keeler . . ."
Quoted in Crowe, Michael, J. *The Extraterrestrial Life Debate, 1750–1900:
The Idea of a Plurality of Worlds from Kant to Lowell.* Cambridge: Cam-
bridge University Press (1986).

56   "'The novelist' was clearly H. G. Wells . . ."
Wells, H. G. *The War of the Worlds.* New York: Berkeley (1964).

57   "As *Viking* project scientist Gerald Soffen . . ."
"Life but no bodies." *New Scientist* (October 14, 1976).

57   "Soffen summed up the general feeling . . ."
Soffen, G. A. "Scientific results of the Viking missions." *Science, 194,*
1274–1276 (1976).

58   "He maintains that the GCMS . . ."
Levin, G. V. "A reappraisal of life on Mars." In *The NASA Mars Conference,*
AAS Science and Technology Series, vol. 71, ed. D. B. Reiber, pp. 187–208
(1988).

58   "In September 2000, a team of scientists . . ."
JPL/NASA press release (September 14, 2000).

59   "In that year a stunning announcement was made . . ."
McKay, D. S., et al. "Search for past life on Mars: Possible relic biogenic
activity in martian meteorite ALH 84001." *Science, 273,* 924–930 (1996).

60   "It was only in 1993, however, that ALH 84001 . . ."
Mittlefehldt, D. W. "ALH 84001, A cumulate orthopyroxenite member of
the martian meteorite clan." *Meteoritics, 29,* 214–221 (1993).

61   "In fact, since 2000, the JSC team has acknowledged . . ."
McKay, D. S. "Evidence for ancient life in Mars meteorites: Lessons
learned." Oral presentation at Workshop on Martian Meteorites, Lunar
and Planetary Institute, Houston, Tex. (November 2–4, 1998).

61   "Again, McKay and his associates have recently shifted their position . . ."
Steele, A., Goddard, D. T., Stapleton, D., Toporski, J. K. W., Peters, V.,
Bassinger, V., Sharples, G., Wynn-Williams, D. D., and McKay, D. S.

"Investigations into an unknown organism on the Martian meteorite Allan Hills 84001." *Meteoritics and Planetary Science, 35,* 273–241 (2000).

62 "A few scientists, including Robert Folk . . ."
Folk, R. L. "SEM imaging of bacteria and nannobacteria [sic] in carbonate sediments and rocks." *Journal of Sedimentary Petrology, 63,* 990–999 (1993)

62 "But, more recently, he and colleague F. Leo Lynch . . ."
Folk, R. L., and Lynch, F. L. "Nannobacteria are alive on Earth as well as Mars." *Proceedings of the International Symposium on Optical Science, Engineering, and Instrumentation (SPIE), 3111,* 406–419 (1997).

63 "Meanwhile, Hojatollah Vali of McGill University . . ."
Vali, H., et al. "Nanoforms: A new type of protein-associated mineralization." *Geochimica et Cosmochimica Acta* (in press).

63 "And, in 1999, they produced new evidence of fossils . . ."
McKay, D. S., et al. "Possible bacteria in Nakhla." Abstract A#1816, *Proceedings of the 30th Lunar and Planetary Conference,* Lunar and Planetary Institute, Houston, Tex. (March 14–19, 1999).

63 "These perfect little six-sided columns, so far as is known . . ."
Thomas-Keprta, K. L., et al. "On the origins of magnetite in Martian meteorite ALH84001." Abstract A#1856, ibid.

64 "Whatever the story is, it's to be published . . ."
Malin, M. C., and Edgett, K. S. "Evidence for recent groundwater seepage and surface runoff on Mars." *Science, 288,* 2330–2335 (2000).

66 "In December 2000 came more extraordinary news."
Malin, M. C., and Edgett, K. S. "Sedimentary rocks of early Mars," *Science, 290,* 1927–1937 (2000).

66 "However, that didn't stop Carl Sagan and his colleague Edwin Salpeter . . ."
Sagan, C., and Salpeter, E. E. "Particles, environments and possible ecologies in the Jovian atmosphere." *Astrophysical Journal Supplement, 32,* 737–755 (1976).

69 "Richard Greenberg and his team . . ."
Greenberg, R., et al. "Tectonic processes on Europa: Tidal stresses, mechanical response, and visible features." *Icarus, 135,* 65–78 (1998).

70 "Perhaps more significantly, as Chris Chyba has pointed out . . ."
Chyba, C. F. "Energy for microbial life on Europa." *Nature, 403,* 381–382 (2000).

70 "Callisto, the outermost of the Galilean satellites, also shows signs . . ."
Khurana, K. K. "Induced magnetic fields as evidence for subsurface oceans in Europa and Callisto." *Nature, 395,* 777–780 (1998).

70 "Infrared observations by the Hubble Space Telescope . . ."
Coustenis, A., et al. "Adaptive optics images of Titan at 1.3 and 1.6 microns

at the CFHT" (in preparation). Results presented at the IAU XXXIVth General Assembly, Manchester, England (August 7–18, 2000).

Chapter 5. Strange New Worlds

75   "On the other hand, if a red dwarf world had a thick atmosphere . . ."
      Joshi, M., Haberle, R., and Reynolds, T. "Simulations of the atmospheres of synchronously rotating terrestrial planets orbiting M dwarfs: Conditions for atmospheric collapse and the implications for habitability." *Icarus, 129*, 450–465 (1997).

76   "An alternative, favored by James Kasting . . ."
      Kasting, J. F. "How Venus lost its oceans." *Oceanus, 32*(2), 54–57 (1989).

78   "But a few years ago, Caltech planetologist David Stevenson . . ."
      Stevenson's ideas concerning the lack of plate tectonics on Venus are discussed by Richard A. Kerr, "The solar system's new diversity." *Science, 265*, 1360–1362 (1994).

79   "Water loss by photodissociation, Kasting found . . ."
      Kasting, J. F. "Habitable zones around stars: An update." In *Circumstellar Habitable Zones*, ed. L. Doyle, Menlo Park, Calif.: Travis House (1996).

80   "However, scientists in the United States and France have recently theorized . . ."
      Forget, F., and Pierrehumbert, R. T.. "Warming Early Mars with carbon dioxide clouds that scatter infrared radiation." *Science, 278*, 1273–1276 (1997).

82   "It goes back to the late 1950s, when the Chinese-American astronomer . . ."
      Huang, S. S. "Occurrence of life outside the solar system." *American Scientist, 47*, 397–403 (1959).

84   "The planets, of which three have been confirmed and a fourth hypothesized . . ."
      Wolszczan, A. "Towards planets around neutron stars." *Astrophysics and Space Science, 212*, 67–75 (1994).

85   "In the end, the honors went to the Swiss pair . . ."
      Mayor, M., and Queloz, D. "A Jupiter-mass companion to a solar-type star." *Nature, 378*, 355-359 (1995).

85   "But no, news soon came of another, similar discovery."
      Butler, P., Marcy, G., Williams, E., Hauser, H., and Shirts, P. "Three new '51 Peg-type' planets." *Astrophysical Journal Letters, 474*, L115 (1997).

86   "That suspicion was quickly backed up by a second discovery . . ."
      Marcy, G. W., and Butler, R. P. "A planetary companion to 70 Virginis." *Astrophysical Journal Letters, 464*, L147–L151 (1996).

87   "Recent theoretical work suggests that the orbits . . ."

Malhotra, Renu. "Migrating planets." *Scientific American 281* (3), 56–63 (September 1999).

88   "Computer simulations show that if two massive planets . . ."
Trilling, D. E., Benz, W., Guillot, T., Lunine, J.I., Hubbard, W. B., and Burrows, A. "Orbital evolution and migration of giant planets: Modeling extrasolar planets." *Astrophysics Journal, 500,* 428–439 (1998).

88   "This could explain a surprising fact that a team of astronomers . . ."
ESO Press Release. "FEROS finds a strange star" (February 2, 1999). Web page: http://www.eso.org/outreach/press-rel/pr-1999/pr-03-99.html

88   "Theorists know of a couple of other ways . . ."
Murray, N., Hansen, B., Holman, M., and Tremaine, S. "Migrating planets." *Science, 279,* 69–71 (1998).

89   "Not only are astronomers adding, almost monthly . . ."
McCaughrean, M. J., and O'Dell, C. R. "Direct imaging of circumstellar disks in the Orion Nebula." *Astronomical Journal, 111,* 1977–1986 (1996).

89   "Caltech's David Stevenson, whom we met earlier . . ."
Stevenson, D. J. "Life-sustaining planets in interstellar space?" *Nature, 400,* 32 (1999).

Chapter 6. Rare Earths and Hidden Agendas

91   "The idea that our planet may be biologically almost unique . . ."
Ward, Peter D., and Brownlee, Donald. *Rare Earth: Why Complex Life is Uncommon in the Universe.* New York: Copernicus (2000).

91   "As its reviewer in the *New York Times* pointed out . . ."
Broad, William A. *New York Times* (February 8, 2000).

91   "*The Times* of London wrote . . ."
The Times (January 26, 2000).

92   "Ward feels that 'an awful lot of astrobiologists . . .' "
Ward quote in Sorensen, Eric. "UW experts squelch hope of finding folks on that final frontier." *Seattle Times* (February 6, 2000).

92   " 'Somehow, I don't think we really appreciated . . .' "
Ibid.

92   "In his review of the book for *Science* . . ."
McKay, C. "All alone after all?" *Science, 288,* 625 (2000).

92   "A similar point is made by the physicist Lawrence Krauss . . ."
Krauss, L. *Physics Today 52*(9), 62 (2000).

92   "As Ward explains: 'I did not go into the debate . . .'"
Private communication.

93   "Worlds and beings of every description were actually predicted . . ."
Dick, Steven J. *Plurality of Worlds: The Extraterrestrial Life Debate from Democritus to Kant.* Cambridge: Cambridge University Press (1982).

94   "Some astronomers, led by Michael Hart . . ."
      Hart, M. H. "The evolution of the atmosphere of the Earth." *Icarus, 33,*
      22–39 (1978).

94   "But a few years later, the pendulum swung back again . . ."
      Kasting, J. F., Whitmire, D. P., and Reynolds, R. T. "Habitable zones around
      main sequence stars." *Icarus 101,* 108–128 (1993).

95   "In 1993, the French astronomer Jacques Laskar . . ."
      Laskar, J., Joutel F., and Robutel, P. "Stabilization of the Earth's obliquity
      by the moon," *Nature, 361,* 615–617 (1993).
      Laskar, J., and Robutel, P. "The chaotic obliquity of planets." *Nature, 361,*
      608–614 (1993).

96   "Guillermo Gonzalez, an astronomer at the University of Washington . . ."
      Gonzalez, G. "Stellar evolution in fast forward." *Astronomy and Geophysics,*
      40, 14–16 (1999).

96   "Recent computer simulations by Eugenio Rivera . . ."
      Rivera, Eugenio J., et al. "Dynamics of the Earth-moon progenitors." Con-
      ference poster session, First Astrobiology Science Conference, NASA
      Ames (April 3–5, 2000).

97   "First, if the Earth had been deprived of a big moon then . . ."
      Shostak, Seth. "Are we so special?" SETI Institute website article (2000).

97   "In 1997, James Kasting and Darren Williams . . ."
      Williams, D. M., Kasting, J. F., and Caldeira, K. "Chaotic obliquity varia-
      tions and planetary habitability." *Circumstellar Habitable Zones,* ed. L.
      Doyle, pp. 43–62. Proceedings of the First International Conference.
      Menlo Park, Calif.: Travis House (1996).

97   "As Alan Boss of the Carnegie Institution explains . . ."
      Pale Blue Dot II conference, Session 5, NASA Ames (May 19–20, 1999).

98   "'Is the lack of a large moon sufficient . . .'"
      Ward, Peter D., and Brownlee, Donald, *Rare Earth,* p. 226.

98   "As Carl Sagan said . . ."
      Davies, Paul. "Survivors from Mars." *New Scientist,* p. 24 (September 12,
      1998).

99   "If the 'Snowball Earth' theory . . ."
      Hoffman, Paul F., and Schrag, Daniel P. "Snowball Earth." *Scientific Amer-*
      ican, 282(1), 68–75 (January 2000).

99   "A much earlier Snowball event, from 2.5 to 2.2 billion years ago . . ."
      Williams, D. M., Kasting, J. F., and Frakes, L. A. "Were low-latitude Pre-
      cambrian glaciations caused by high obliquity?" *Nature, 396,* 453–455 (1998).

99   "And according to a recent study . . ."
      Culler, T. S., Becker, T. A., Muller, R. A., and Renne, P. R. "Lunar impact
      history from $^{40}Ar/^{39}Ar$ dating of glass spherules." *Science, 287,* 1785–1788
      (2000).

100   "As Athena Andreadis, a neurologist . . ."
SETI League on-line review of *Rare Earth* by Athena Andreadis, "E.T., call Springer-Verlag!" at http://www.setileague.org/articles/rarearth.htm.

101   "Yet they go on to endorse the idea . . ."
Ward, Peter D., and Brownlee, Donald, *Rare Earth*, pp. 113–124.

101   "They even point to a possible further trigger . . ."
Kirschvink, J. L., Ripperdan, R. L., and Evans, D. A. "Evidence for a large-scale Early Cambrian reorganization of continental masses by inertial interchange true polar wander." *Science, 227*, 541–545 (1997).

102   "NASA Ames astrobiologist Chris McKay . . ."
McKay, C. "Time for intelligence on other planets." In *Circumstellar Habitable Zones*, ed. L. Doyle, pp. 405–419. Menlo Park, Calif.: Travis House (1996).

102   "Ward and Brownlee counter . . ."
Ward, Peter D., and Brownlee, Donald, *Rare Earth*, p. 219.

102   "This passage is typical of the extent . . ."
Ward, Peter D., and Brownlee, Donald, *Rare Earth*, p. 147.

105   "And Debra Fischer at the University of California, Berkeley . . ."
Press release at International Astronomical Union XXIVth General Assembly, Manchester, England, August 7–18, 2000. Data unpublished.

105   "In October 2000, George Gatewood . . ."
Prepublication announcement by George Gatewood, Inwoo Han, and David C. Black at Division for Planetary Sciences of the American Astronomical Society meeting, Pasadena, Calif. (October 23–27, 2000).

107   "This idea was originally put forward by George Wetherill . . ."
Wetherill, G. W. "How special is Jupiter?" *Nature, 373*, 470 (1995).

108   "Guillermo Gonzalez has been especially vocal . . ."
Gonzalez, G. "Spectroscopic analysis of the parent stars of extrasolar planetary systems." *Astronomy & Astrophysics, 334*, 221–238 (1998).

108   "HD 37124 is orbited by an eccentric Jupiter . . ."
Vogt, S. S., Marcy, G. W., Butler, R. P., and Apps, K. "Six new planets from the Keck Precision Velocity Survey." *Astrophysical Journal, 536*, 902–914 (2000).

109   "This, again, is a concept that Guillermo Gonzalez . . ."
Guillermo, Gonzalez, and Ward, Peter. "The galactic habitable zone." Conference poster session, First Astrobiology Science Conference, NASA Ames (April 3–5, 2000).

111   "Alan Rubin, for example, produced a skeptical piece . . ."
Rubin, A. "Paucity of aliens." *Griffith Observer, 61*, 2–14 (1997).

111   "Although he personally may not be well known . . ."
Gonzalez, Guillermo. "Nobody here but us Earthlings." *Wall Street Journal* (July 16, 1997).

111   "'It's courageous,' comments Marcy."
Broad, William A. *New York Times* (February 8, 2000).

111   "But Marcy, like the great majority of astronomers . . ."
Marcy quoted in "Finding worlds like our own," University of California-Berkeley news release (October 30, 2000).

111   "As Ward and Brownlee point out in their preface . . ."
Ward, Peter D., and Brownlee, Donald, *Rare Earth*, p. x.

111   " 'We often met,' recalls Gonzalez . . ."
Private communication.

112   " 'It was not something I took lightly,' he explains . . ."
Private communication.

112   " 'Live Here or Nowhere' concludes . . ."
Ross, Hugh, and Gonzalez, Guillermo. "Live here or nowhere." *Connections*, *1*(4) (1999).

112   "One of the references in this article . . ."
Gonzalez, G. "Is the sun anomalous?" *Astronomy & Geophysics*, *40*(5) 25–29 (1999).

113   "In a 1997 piece, he writes . . ."
Gonzalez, Guillermo. "Minicomets write new chapter in earth-science." *Facts and Faith*, *11*(3) (1997).

113   "A 1998 article, 'Design Update: How Wide is the Life Zone?" . . .' "
Gonzalez, Guillermo. "Design update: How wide the life zone?" *Facts & Faith*, *12*(4) (1998).

113   "'Gonzalez has been a big influence,' commented Ward."
Private communication.

113   "But an inquiry to Ward on this issue . . ."
Private communications.

114   "Copernicus, Einstein, and the modern cosmologists . . ."
Greene, Brian. *The Elegant Universe: Superstrings, Hidden Dimensions and the Quest for the Ultimate Theory.* New York: W. W. Norton (1999).

114   "Agassiz, who had arrived in the United States . . ."
Larson, Edward J. *Inspired by the World's End: A History of Science on the Galapagos.* New York: Basic Books (2001).

Chapter 7. Theme and Variation

117   "The physicist Guiseppe Cocconi . . ."
Swift, David. *SETI Pioneers: Scientists Talk About Their Search for Extraterrestrial Intelligence.* Tucson: University of Arizona Press (1990).

119   "For the simple reason that Darwin wrote a book . . ."
Darwin, Charles. *The Origin of Species.* New York: Grammercy (1998). First published in 1857.

120   "In *The Time Machine* ..."
       Wells, H. G. *The Time Machine*. New York: Dover (1995). First published in book form in 1885.

120   "Then, in April 1896, in a Saturday Review article ..."
       Wells, H. G. "Intelligence on Mars." *Saturday Review, 81,* 345–6 (April 4, 1896).

120   "A year later, in *The War of the Worlds* ..."
       Wells. H. G. *The War of the Worlds*. New York: Berkeley (1964).

121   "He found it 'noticeable that, as usual ...' "
       Matthew, W. D. "Life on other worlds." *Science, 44,* 239 (1921).

122   "In their hugely popular 1931 encyclopedia ..."
       Wells, H. G., Huxley, Julian, and Wells, G. P. "Is there extra-terrrestrial Life?" In *The Science of Life*, ed. H. G. Wells, Julian Huxley, and G. P. Wells, p. 11, New York: Doubleday (1931).

122   "Twenty years later, Harold Blum ..."
       Blum, Harold. *Time's Arrow and Evolution*. 3rd ed. Princeton, N.J.: Princeton University Press (1968). First published in 1951.

122   "In his 1989 book *Wonderful Life* ..."
       Gould, Steven Jay. *Wonderful Life: The Burgess Shale and the Nature of History*. New York: W. W. Norton (1989).

124   "In his 1998 book *The Crucible of Creation* ..."
       Conway Morris, Simon. *The Crucible of Creation*. New York: Oxford University Press (1998).

124   "In his great work *On Growth and Form* ..."
       Thompson, D'Arcy Wentworth, *On Growth and Form*.

125   "Eyes, for instance, have been discovered ..."
       As estimated in Dawkins, Richard. *River Out of Eden*. New York: Basic Books (1996).

125   "And speaking of animals that swim ..."
       Conway Morris, Simon, *Crucible of Creation*.

125   "So-called 'structural convergence' seems to be the only way ..."
       Roux, K. H., et al. "Structural analysis of the nurse shark (new) antigen receptor (NAR): Molecular convergence of NAR and unusual mammalian immunoglobulins." *Proceedings of the National Academy of Sciences USA, 95,* 11804–1809 (1998).

126   "Cheng Chi-hing and her co-workers ..."
       Chen, L., DeVries, A. L., and Cheng, C.-H. "Evolution of antifreeze protein from a trypsinogen gene in Antarctic notothenioid fish." *Proceedings of the National Academy of Sciences USA, 94,* 3811–3816 (1997).

127   "As Conway Morris puts it ..."
       Conway Morris, Simon, *Crucible of Creation*.

128   "An experiment done by Andrew Ellington and his colleagues ..."

Described in Hesman, T. "Code breakers: Scientists are altering bacteria in a most fundamental way." *Science News, 157*(23), 360 (June 3, 2000).

129   "The recent discovery of a simple sugar molecule . . ."
Hollis, J. M., Lovas, F. J., and Jewell, P. R. "Interstellar glycolaldehyde: The first sugar." *Astrophysical Journal Letters, 540,* L107–L110 (2000).

130   " 'When we get to some other planet . . .' "
Morowitz quote in Holmes, R. "Life is . . ." *New Scientist,* p. 40 (June 13, 1998). See also: Morowitz, H. J., Kostelnik, J. D., Yang, J., and Cody, G. D. "The origin of intermediary metabolism." *Proceedings of the National Academy of Sciences USA, 97,* 7704–7708 (2000).

130   "To back up this claim, Morowitz and his colleagues in 1998 . . ."
Morowitz, H. J., et al. "The origin of intermediatry metabolism," *Proceedings of the National Academy of Sciences, 97*(14), 7704–7708 (2000).

130   "As the biochemist George Wald pointed out . . ."
Wald, G. "Life and light." *Scientific American* (October 1959).

134   "As James Shapiro, a microbiologist . . ."
Shapiro, James A. "Bacteria as multicellular organisms." *Scientific American, 258*(6), 62–69 (June 1988).

136   "But in 1999, Jochen Brocks . . ."
Brocks, J. J. "Archaean molecular fossils and the early rise of eukaryotes." *Science, 285,* 1033–1036 (1999).

136   "Kirschvink's proposal was used by Ward and Brownlee . . ."
Ward, Peter D. and Brownlee, Donald, *Rare Earth,* pp. 93–94.

137   "A way out this enigma has been suggested . . ."
Knoll, A. H. "A new molecular window on early life," *Science, 285,* 1025–1026 (1999).

137   "As Lynn Margulis and her son Dorian Sagan wrote . . ."
Margulis, Lynn, and Sagan, Dorion. "The parts: Power to the protoctists." *Earthwatch* (September/October 1992).

138   "Martin Boraas and his colleagues . . ."
Boraas, M. E., Seale, D. B., and Boxhorn, J. E. "Phagotrophy by a lagellate selects for colonial prey: A possible origin of multicellularity." *Evolutionary Ecology, 19,* 153–164 (1998).

139   "As James Shapiro explains . . ."
Shapiro, James A., "Bacteria as multicellular organisms."

139   "Using strains of *Escherichia coli* whose pedigrees . . ."
Adler, J., and Tso, W.-W. " 'Decision'-making in bacteria: Chemotactic response of *Escherichia coli* to conflicting stimuli." *Science, 184,* 1292–1294 (1974).

139   "Daniel Koshland, of the University of California, Berkeley . . ."
Koshland, Daniel E., Jr. "A response regulator model in a simple sensory system." *Science, 196,* 1055–1063 (1977).

140   "Toshiyuki Nakagaki and his colleagues . . ."
       Nakagaki, T., Yamada, H., and T A'oth, A'A. "Intelligence: maze-solviong
       by an amoeboid organism," *Nature, 407,* 493–496 (2000).

141   "A recent analysis by Charles Lineweaver, . . ."
       Lineweaver C. H. "An estimate of the age distribution of terrestrial planets
       in the universe: Quantifying metallicity as a selection effect," *Icarus* (sub-
       mitted September 2000).

142   "Gavin Hunt, of Massey University in New Zealand . . ."
       Hunt, G. R. "Manufacture and use of tools by New Caledonian crows."
       Nature, 379, 249–251 (1996).

142   "Bernd Heinrich of the University of Vermont . . ."
       Heinrich, Bernd. *Ravens in Winter.* New York: Vintage (1991).

142   "Irene Pepperberg, at the University of Arizona . . ."
       Pepperberg, I. M. "Cognition in the African Grey Parrot: Preliminary evi-
       dence for auditory/vocal comprehension of the class concept." *Animal
       Learning and Behavior, 11*(2), 179–185 (1983).

143   "The female lowland gorilla, Koko . . ."
       Patterson, F. G. P., and Cohn, R. H. "Language acquisition by a lowland
       gorilla: Koko's first ten years of vocabulary development." *Word, 41*(2),
       97–144, 1991.

143   "From its remains, Dale Russell . . ."
       Russell, D. A., and Sequin, R. "Reconstruction of the small Cretaceous
       theropod *Stenonychosaurus inequalis* and a hypothetical dinosauroid." *Syl-
       logeous, 37*(1) (1982).

Chapter 8. Life Signs

145   "One of the scientists involved . . ."
       Horowitz, Norman H. To Utopia and Back: The Search for Life in the Solar
       System. New York: W. H. Freeman (1987).
       Levin, G. V. "A reappraisal of life on Mars." In *The NASA Mars Conference,*
       AAS Science and Technology Series, vol. 71, ed. D. B. Reiber, pp. 187–208
       (1988).

146   "As Harvard biologist Andrew Knoll has pointed out . . ."
       John Kerridge spoke of "Knoll's Rule" at the National Academies Space
       Studies Board Life Detection Techniques Workshop at the Carnegie Insti-
       tution (April 25–27, 2000).

146   "According to Lovelock . . ."
       Lovelock, James. "Gaia: The world as a living organism." *New Scientist,*
       p. 28 (December 18, 1988)

146   "James Kirchner, a geologist at the University of California . . ."
       Kirchner, J. W. "The Gaia Hypotheses: Are they testable? Are they useful?

In *Scientists on Gaia,* ed. S. Schneider. Cambridge, Mass.: MIT Press (1991).

147 "In fact, measurements by the *Galileo* probe . . ."
Sagan, C., et al. "A search for life on Earth from the *Galileo* spacecraft." *Nature, 365,* 715–721 (1993).

148 " 'Life is a phenomenon that exists . . .' "
Lovelock, James. "Gaia: The world as a living organism." *New Scientist* pp. 25–28 (December 18, 1986).

148 "Those hopes were given a boost . . ."
Kral, Timothy, and Bekkum, Carl. "Growth of a methanogen on Mars soil simulant in a water-limited environment." 99th General Meeting of the American Society for Microbiology, Chicago, Ill. (May 30–June 3, 1999).

149 "The Idaho group is developing a device . . ."
University of Idaho news release, "UI team wins grant to help NASA detect life on other planets" (January 7, 2000).

150 "Recent evidence for liquid water near the surface of Mars . . ."
Malin, M. C., and Edgett, K. S., "Evidence for recent groundwater seepage."

150 "Scientists recently managed to revitalize spores . . ."
Described by Travis, J. "Prehistoric bacteria revived from buried salt." *Science News, 155,* 373 (June 12, 1999).

152 "If you talk to Caltech's Joseph Kirschvink . . ."
Kirschvink, J., and Vali, H. "Criteria for the identification of bacterial magnetofossils on Earth or Mars." Abstract A#1681, *Proceedings of the 30th Lunar and Planetary Science Conference,* Lunar and Planetary Institute, Houston, Tex. (March 14–19, 1999).
Thomas-Keprta, K., et al. "On the origins of magnetite."

152 "A now-retired meteorite specialist . . ."
Kerridge, J. "Martian exobiology in the Post-ALH 84001 era: Some key issues." Oral presentation, 60th Annual Meteoritical Society Meeting, University of Hawaii (July 21–25, 1997).

152 "That would tie in with a suggestion by Christopher McKay and . . ."
Hartman, H., and McKay, C. P. "Oxygenic photosynthesis and the oxidation state of Mars." *Planetary and Space Science, 43,* 123–128 (1995).

153 "As John Kerridge says . . ."
Kerridge quote from the National Academies Space Studies Board, Life Detection Techniques Workshop at the Carnegie Institution (April 25–27, 2000).

153 "This possibility was given a further boost in October 2000 . . ."
Weiss, B. P., Kirschvink, J. L., Baudenbacher, F. J., Vali, H., Peters, N. T., Macdonald, F. A., and Wikswo, J. P. "Reconciliation of magnetic and pet-

rographic constraints on ALH84001? Panspermia lives on!" Abstract A#2078, 31st Lunar and Planetary Science Conference, NASA Johnson Space Center, Houston, Tex. (March 13–17, 2000).

156   " 'The goal of both rovers will be to learn about ancient water and climate . . .' "
NASA HQ Press Release 00–124, "Twin rovers headed for Mars."

159   "Fortunately, we have a remarkable analogue . . ."
Siegert, Martin, J. "Antarctica's Lake Vostok." *American Scientist, 87*(6) 510–517 (1999).

162   "Subsequent analysis at Purdue University . . ."
Lipschutz, Michael, and Friedrich, Jon. Presentations on laboratory analysis of Tagish Lake meteorite, 63rd Annual Meeting of the Meteoritical Society, Chicago, Ill. (August 23–September 1, 2000).

162   "Mission scientists reported in 2000 . . ."
"Tarlike macromolecules detected in 'stardust'," Max Planck Society Research News Release (April 26, 2000).

164   " 'We're now at a stage,' explains extrasolar planet hunter Geoff Marcy . . ."
Marcy comment at International Astronomical Union XXIVth General Assembly, Manchester, England (August 7–18, 2000).

164   "One of these extra blips spotted in 1998 . . ."
Rhie, S. H., et al. "On planetary companions to the MACHO 98-BLG-35 microlens star." *Astrophysical Journal, 533,* 378–391 (2000).

Chapter 9. The Cosmic Community

174   "He believes that 'subsurface life may be widespread . . .' "
Gold, Thomas. *The Deep, Hot Biosphere.* New York: Springer-Verlag (1999).

175   "As he points out, 'one may even speculate that . . .' "
Ibid.

176   "The best place to find pristine examples of biological material . . ."
Shklovskii, I. S., and Sagan, Carl. *Intelligent Life in the Universe.* San Francisco: Holden-Day (1966).

177   "It was first discussed in 1961 . . ."
Barrow, John D., and Tipler, Frank. *The Anthropic Cosmological Principle.* New York: Oxford University Press (1986).

Index